T0300477

Investment Grade Energy Audit:

Making Smart Energy Choices

Investment Grade Energy Audit:

Making Smart Energy Choices

Shirley J. Hansen, Ph.D.
James W. Brown, P.E.

THE FAIRMONT PRESS, INC.
Lilburn, Georgia

MARCEL DEKKER, INC.
New York and Basel

Library of Congress Cataloging-in-Publication Data

Hansen, Shirley 1928-
  Investment grade energy audit : making smart energy choices / Shirley
J. Hansen, James W. Brown.
    p. cm.
  Includes bibliographical references and index.
  ISBN 0-88173-362-8 (print) -- ISBN 0-88173-464-0 (electronic)
  1. Energy conservation. 2. Energy policy. I. Brown, James W. II. Title

  TJ163.3.H363 2004
  658.2'6--dc22

2003056515

Investment grade energy audit : making smart energy choices / Shirley J. Hansen, James W. Brown.

Cartoons ©2003 by Stephen C. Hansen. Used with permission.

Published by The Fairmont Press, Inc.
700 Indian Trail, Lilburn, GA 30047
tel: 770-925-9388; fax: 770-381-9865
http://www.fairmontpress.com

Distributed by Marcel Dekker, Inc.
270 Madison Avenue, New York, NY 10016
tel: 212-696-9000; fax: 212-685-4540
http://www.dekker.com

Printed in the United States of America

10   9   8   7   6   5   4   3   2   1

0-88173-362-8 (The Fairmont Press, Inc.)
0-8247-0928-4 (Marcel Dekker, Inc.)

Table of Contents

Foreword

Years ago there was a TV show called "What's My Line?" where celebrities would ask questions of several contestants to determine which participant actually had the unusual occupation described by the moderator. After the celebrities guessed which contestant actually held the position, the moderator's tag line was, "Will the real _____ please stand up?"

Since we started using the term, *Investment Grade Audit*, several years ago to denote a higher level of auditing, many have adopted the name to describe the traditional audit they still use. Far too many have changed the name, but not the game.

Well, "investment grade" does have a nice ring to it, doesn't it? Unfortunately, we have not been able to come up with another label that better describes the audit that captures the sense of an investment guide. So we are left with asking, "Will the *real* Investment Grade Audit please stand up?"

What questions can our celebrities (the customers, consulting engineers, and financiers) ask to help determine if a real Investment Grade Audit (IGA) has been conducted?

That's what this book is about: How to determine if you have been presented with a real IGA or just the same ol' stuff with a new wrapping. Or, if you are an engineer or an ESCO, how you can be satisfied that you are truly offering an IGA. So we discuss why the old traditional audit is no longer good enough and what the auditor must do to raise the bar in auditing practices to deliver an IGA.

We need to clear the air regarding that fuzzy crystal ball often referred to as "predicted energy savings." It is the basic underpinning of the energy efficiency industry. An industry that rests heavily on trust. An industry, which regularly talks about this nebulous idea of reducing hot air. Talk about trust!

To get from "here" to "there," we need to: 1) examine the predictive consistency of the auditor; 2) whether the audit addressed the broad implications of energy efficiency; not just conservation; 3) if the audit considered the full ramifications throughout the organization's operation of the contemplated energy efficiency measures, and 4) if the auditor has considered how those measures will behave over time.

We do a tremendous disservice to ourselves, the industry, our customers, and to those who depend on us for all the potential economic and environmental benefits, to deliberately mislead people as to the quality of the audit being delivered. Just giving the same old stuff a new name does not do the job. We sincerely hope this book will make those who offer, and those who accept, "IGAs" think twice about what they are doing. Only then will the Investment Grade Audit be able to "stand up" as the real thing. Only then will it have the meaning originally intended throughout the industry and to its customers.

Preface

What a delightful world it would be if one could envision a book and suddenly it was there. Unfortunately, it does not happen that way. Many laborious hours of research, writing and rewriting are invested in an effort such as this. And through it all, many people aid and abet.

I would first like to thank my co-author, James W. Brown, president of Energy Systems Associates. Jim brings to this endeavor two great assets: 1) he is that wonderful, all too rare, creature known as an articulate engineer; and 2) he has lived through the rigors of analyzing where the traditional audits were falling short and breaking new ground as to what an investment grade audit should be. Today, his firm can document the value of offering its clients an audit that stands the test of time. We are indebted to Jim for sharing some of his firm's very valuable proprietary information.

I say "we" because I also speak for my business associate, colleague and husband, James C. Hansen, who helped write much of the book, but refused to have his name on the cover. Jim B. and I thank you for your incredible contributions and your patience.

Jim B. declares that engineers like to look at pictures; so we have him to thank for endorsing the idea of illustrations to break up all that print. Once again, we are very pleased to share some of Stephen Hansen's unique perspective of the world with our readers. While rushing to get work ready for his one-man show in Washington, DC, he found time to read the manuscript and find the humor in the perplexities of auditing. Our apologies to Stephen and his manager for scuttling his time schedule—as only a mother can do. And our sincere thanks to him for giving us some rather whimsical characters that help to highlight some of the points we try to make.

Finally, we'd like to thank the very patient and delightful people at The Fairmont Press for putting up with our delays and difficulties in getting the job done.

We all hope that this book leads all its readers to make "Smart Energy Choices."

Shirley J. Hansen

Chapter 1

How Auditing Evolved

ONCE UPON A TIME, a visionary named Edward Stephen encouraged an engineer named Goody Taylor to survey four schools in Fairfax County, Virginia, to see if there might be some way for the school district to cut its utility bills. The year was 1962. The term energy audit had yet to be invented but the thought had been born. Cheap and available energy, however, caused this strange, innovative idea to languish for more than a decade.

Forty years ago most of us thought of "energy" as something associated with "kinetic" in a physics class. The electricity and fuel we used to run things was really cheap and we all expected to be able to pump whatever we needed out of the ground... forever. We were so parochial that the average American thought of the Middle East as a place where men ran around with tablecloths draped over their heads—if we thought of it at all. We had little interest, and certainly no "vital interest," in what happened there.

Few facility managers, at the time, had any idea as to what types of equipment could, and should, be operated more efficiently. And it would have cost more for the Goody Taylors of that era to survey our buildings than we probably would have recovered in utility savings.

Ed Stephen was indeed a man before his time, but the times caught up with him. Not surprisingly, when the need arose he took the lead at the Federal Energy Administration (FEA, a precursor of the U.S. Department of Energy [DOE]) to create an auditing protocol.

When the "energy crisis" of the 1970s struck, it quickly became apparent that engineers could not, with any degree of confidence, accurately project energy savings. There followed a

1

struggle to develop a uniform protocol to assess a range of energy conservation measures in a variety of building types. Ed Stephen stepped forward again, using FEA funding to support the Saving Schoolhouse Energy project. This project provided funds for 11 engineers from around the country to get together to establish an energy survey protocol. The implementation of their recommendations demonstrated that engineers could predict energy savings if they needed to. They had just never needed to before!

EARLY AUDITING STRUGGLES

Did you ever wonder how the term "audit" became associated with an energy efficiency survey? In the 1977 Federal Register,[1-1] the federal government offered its view, calling this survey an "audit," to be performed by an "auditor"—*a certified public accountant*! Those of us who were scratching our heads trying to determine how we could establish a methodology for figuring potential energy savings in mechanical systems and building shells viewed the whole idea of a CPA doing an energy audit with considerable amusement.

Now, we really hate to admit it, but 20/20 hindsight would suggest that there was an element of logic in what the feds did. While shaking our heads at the feds, we let the pendulum swing all the way to the technological side and unfortunately, ignored some important financial considerations.

By the late 1970s and early 1980s, auditing became a battlefield between consulting engineers and energy service companies (ESCOs). Consulting engineers frequently had a long established relationship with a client and did not take kindly to ESCOs getting in the middle of this relationship. And sometimes discrediting the engineer's work. The engineers often asserted that they best represented the clients' interests and would protect them from the ESCOs' "capitalist greed." After all, wasn't it obvious that the ESCO was there to invade the client's domain and make money off their operation?

The ESCOs retaliated by pointing out that the engineers gave clients audit reports and collected a fee—and were not held accountable for their savings projections. These engineers, the ESCOs pointed out, usually received a fee as a percentage of the size of the job. The bigger the job; the bigger the fee. It's this kind of thinking that led to designing and installing three oversized boilers for a modest elementary school in Maryland. Such over design in a relatively mild climate can only be explained one way. How did this, ESCOs queried, represent the client's interests?

There was no question that the ESCO movement threatened traditional engineering turf, at a time when the whole business of auditing in the United States was less than ten years old. In addition, ESCO demands for engineering accountability brought new pressure on the technical capabilities of the auditors.

On the plus side, the ESCO industry raised auditing expectations. If ESCOs were to bet money on audit projections, a better quality of audit was definitely needed.

Among these cross-currents, the concept of "shared savings" grew and ESCOs began using audits as marketing tools to convince owners of the savings potential in their facilities. Unfortunately, some ESCO sales engineers, in their zeal to sell a project, over-stated the savings potential. Most ESCOs at the time paid their sales people in commissions at contract signing. This inherent conflict of interest prompted some to "cook the books" by exaggerating the savings. Only much later, did the ESCOs realize it was in their interest to pay most of the sales commission *after* the projects had proven themselves.

In a diabolical twist, some ESCO auditors deliberately "lowballed" the potential savings estimates. Since owners had agreed to pay the ESCOs a percentage of the cost savings, higher savings than predicted meant ESCOs got more than the owners had expected to pay for the installed equipment. For example, owners were lulled into seeing 70 percent of the projected one million dollars/year in savings on a five-year contract as reasonable for $3.5 million in equipment. Instead of one million per year, however, the project saved two million per year—and the equipment "price tag" doubled. As a result, all too often customers paid as much as $7 million for $3.5 million worth of equipment

It is no wonder that a few such deals created great consumer skepticism. Engineers, still trying to protect their turf, were quick to add fuel to this uncertainty.

Also stepping into the breach, or the confusion, was the advent of utilities attempting to do residential audits at the behest of the federal government. Where else but the good ol' US of A would the federal government turn to an industry that had the least to gain from helping consumer use less energy... companies that were in the business of selling energy? The feds managed to do a thorough job of stirring the auditing pot by requiring utilities under the Residential Conservation Service to perform audits.

Like any industry, utilities had their own perspectives and expertise. The Tennessee Valley Authority (TVA), for example, was offering free audits. But the TVA had the Tennessee Energy Office wringing its hands because the TVA audits only looked at potential electricity savings. So in addition to the auditing protocols evolving out of DOE's Institutional Conservation Program for schools and hospitals, another portion of DOE was helping utilities learn how to audit residences. (Not to be out done, those in DOE, who were working with industry were also trying to develop audit guidance.)

It's a wonder that auditing survived and prospered during these early years, but it did!

By the late 1970s, at the first World Energy Engineering Congress, Mr. Nick Choksi took one of the first stabs at describing an energy audit, saying that audits "...require complex and thorough research by knowledgeable and competent people to provide practical and realistic results." Even more true today! Choksi went on to say that his Certified Test and Balance Company had "established five major steps to prepare an integrated energy audit.

1. Energy inventory   4. Analysis
2. Engineering overview   5. Final report"[1-2]
3. Data collection

Those steps have not changed much over the years, but the substance within these steps has. Since that time, we have learned much about energy conservation and efficiency. We know window treatment's cost-effectiveness varies with which side of the building is being treated. We learned that the needed "U" value of roof insulation should take into consideration the local climate and the existing components of the roof. And we certainly, although gradually, did a much better job of determining what equipment should be installed, what shell modifications should be made, and roughly how much energy we could be expected to save.

We also learned about the work environment, human behavior and productivity. But the batting record isn't so good on those scores, as little of this wisdom has found its way into energy auditing practices.

GAINING CONFIDENCE IN THE '80s

By the mid-1980s, we were at a point where we rather confidently assigned paybacks to certain types of measures. We "knew" that lighting had a four year payback, building shell work ranged from 5 years for storm windows to 8 years for such things as insulation. DOE issued the following bar chart that seemed to make it much easier for engineers to determine what the payback was going to be for a set of measures—a mistaken assumption. The focus, unfortunately, was entirely on how quickly certain types of measures could pay for themselves in savings.

Since the early 1980s, we have basically fine-tuned the technical aspects of the audit, and then fine-tuned them even further. But beyond these technical refinements other factors have all too frequently received a passing nod at best.

Having helped give birth to the energy audit born in the late 70s, the authors can say with some pride that, for the most part, the traditional energy audit has served us well. But as Will Rodgers is credited with saying, "Even if you are on the right road; if you just sit there, you are going to get run over."

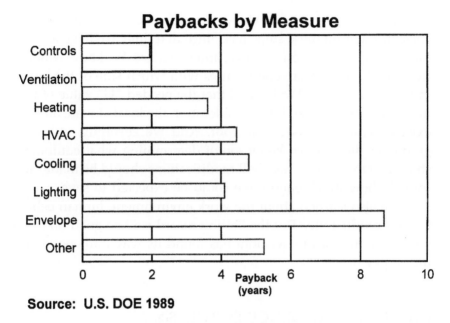

Source: U.S. DOE 1989

Figure 1-1. Paybacks by Measure

THE ENERGY AUDIT

The energy audit most in use today generally attributes its roots to the auditing protocol developed for use under DOE's Institutional Conservation Program (ICP). Through the years, the audit has become more sophisticated and more inclusive, but until recently it retained its "snap shot" characteristics, making predictive consistency challenging at best. We use the term, "snap shot" to focus on the typical assumption auditors made, and continue to make, that the facility or process conditions, as they appear during the audit, will remain the same for the life of the equipment or project.

As we have grown in sophistication, a general consensus has emerged that those buildings and processes probably don't stay the same, but we have not systematically figured a way to quantify what changes might take place, nor the degree to which those changes might impact the projected savings. This book is designed to build on our strong technical capabilities and guide the auditor in weighing factors and conditions that will impact on equipment performance, overall project performance, and projected savings.[1-3]

TYPES OF AUDITS

There has emerged two relatively discrete types of audits that might be performed on a facility/process; i.e., the scoping audit and the engineering feasibility study. In practice, especially from the owner's perspective, they may be blended into one.

The "Scoping" Audit

Sometimes the scoping audit is not much more than a "walk through," a cursory examination of a facility and its energy using systems. This sounds more casual than it is (should be) because preliminary data is gathered which will help to determine whether the potential for an economically viable project exists. This preliminary work will determine whether a much more expensive, much more detailed "engineering feasibility" study is warranted. The scoping audit generally will include a basic description of the facility, its function, information on energy using systems and gross energy use.

In recent years, ESCOs have also found that at least a preliminary exploration as to the creditworthiness of the facility owners can a very valuable part of this scoping audit. Knowing whether or not a project can be financed by the client before serious money is spent on detailed engineering studies can save time, money and embarrassment.

Sketches are usually made showing the general layout of the facility (at this point, it may be little more than a fire escape diagram) and information on square footage is added. A simple descriptive inventory of energy using systems and equipment is assembled along with name plate data, age and notes on condition. Utility and other energy bills are reviewed and, where appropriate, energy indices are calculated. Quick "back of the envelope" calculations are made that suggest possible energy savings. These are preliminary calculations and measurements, designed primarily to provide enough information to determine whether a detailed engineering audit is justified. The scoping audit costs money, but it is "cheap insurance" when compared to the cost of an engineering feasibility study that wasn't warranted.

If after a review of the preliminary data, it appears that an opportunity for a viable project exists, and the potential client agrees, a detailed engineering feasibility study gets underway.

The Engineering Feasibility Study

This study builds on the preliminary data, substituting measurements for estimates, verification for approximation. (Under DOE's Institutional Conservation Program, this level of audit was referred to as a Technical Assistance analysis, or TA.) The study examines facility use and patterns to determine function, occupant loads and timing and all of the other factors that influence energy usage. The building shell is evaluated to determine points of energy loss or opportunities for improvements. Lighting is examined with particular care—and a good auditor *never* takes the word of building occupants or facility managers as to run time—actual measurements are almost always justified. [The use of a Portable Data Logger, described in Chapter 4, "Building the M&V Foundation," is an inexpensive way to establish operating hours.] Heating, cooling and other equipment is evaluated and required temperature, humidity, air exchange rates and other environmental parameters are carefully noted. All energy using equipment is inventoried, its condition noted along with hours of operation and energy required. We refer to this type of audit as the "traditional" audit.

Utility bills are examined (covering at least one full year) with attention to demand charges and load profiles. Rate schedules are checked to be sure the correct rates are being charged (applying the correct rate schedules are often a source of impressive "energy savings"). In some instances, weather statistics for the same time period may be important if variations from normal weather patterns have been sufficient to have a significant bearing on energy use and the measures under consideration are temperature dependent.

Careful attention is paid to operations and maintenance (O&M) of the equipment, which impacts energy consumption. The O&M assessment should give special attention to the skill levels of O&M personnel. This information becomes the basis for determining what equipment should be recommended and will provide an indication of the amount of training that must be supplied. The objective of this training should be to assure that any new equipment will operate near design and projected energy savings will be achieved throughout the life of the project.

Equipment additions, replacements, modifications or improvements are recommended, generally with detailed specifications. Based upon calculated payback periods, equipment selections are made with the expectation that certain energy efficiency improvements will be achieved.

More recently a detailed examination of the ownership and financial solvency of the facility owners has been incorporated into the technical feasibility study... a move toward the *investment grade audit*. Some financial institutions now provide a checklist of information about facility ownership and financial health, which must be completed before they will begin to look at a proposed financial package for an energy efficiency project. As the pendulum swings back toward investment criteria, the cost-effectiveness of specific measures becomes more critical.

COST-EFFECTIVENESS

In 1983, Al Thumann, in his *Handbook of Energy Audits*, excerpted a quote from the first generation of Class C^{1-3} workbooks,

> The energy audit serves to identify all of the energy streams into a facility and to quantify energy use according to discrete functions.

An energy audit may be considered as similar to the monthly closing statement of an accounting system. One series of entries consists of amounts of energy, which were consumed during the month in the form of electricity, gas, fuel oil, steam, and the second series lists how the energy was used; how much for lighting, air conditioning, heating, process, etc. The energy audit process must be carried out accurately enough to identify and qualify the energy and cost savings that are likely to be realized through investment in an energy savings measure.[1-4]

Before we engage further in this discussion, it's important to recognize all the valuable guidance Al Thumann and others have offered to improve the technical competence of auditors. This book is designed to augment, not replace, such excellent counsel. Nowhere do we discuss the relative merits of installing "widgets" or "gismos" to save energy or money. The authors assume that the reader has a full command of traditional auditing techniques, but has an intellectual thirst for enhancing those capabilities. Throughout this book every effort is made to build on current auditing techniques and to raise the bar on the quality of the report an auditor offers a client. In turn, we hope the client will be better prepared to insist upon, even demand, audit recommendations that provide true investment guidance.

Even though we didn't fully appreciate it at the time, the above quotes from the Class C workbooks were on target by focusing on the investment aspect. Fundamental to any quality audit is the premise: *Energy efficiency is an investment; not an expense.*

The building stock and industrial processes are major portions of the owner's investment portfolio. An energy audit should be more than a guide to saving energy; it should be an investment guide for enhancing that investment portfolio. The cost/benefit analysis presented in the audit report should offer the investor a reliable guide to the investment potential of the recommended measures—a guide that recognizes the impact to the organization's bottom line. All too many traditional audits have fallen miserably short of this goal.

No matter how many times we used the term, *cost-effective*, and declared "Energy Conservation makes money!!" we didn't quite get it. In fact, in retrospect, it's really quite easy to see why bankers did not rush to finance our early "energy conservation measures." It was not possible to "bank on" the predicted paybacks or related financial calculations.

For a banker, simple interest at 10 percent provides a 100% return on investment (ROI) in 10 years. But a four-year payback in predicted energy savings did not offer a clear certainty that the investment would truly be returned in four years. A two-year payback should offer a 50 percent ROI; a three-year payback; 33 percent, etc. Unfortunately, so many things could, and did, happen to affect the energy saving revenue streams that those predictions were sometimes off by as much as 400 percent—an unacceptable risk. Not surprisingly, early efforts to finance energy efficiency were typically confined to the firm's balance sheet or the value of the installed equipment as collateral; promised savings did not serve.

In retrospect, one can't help wonder if we used "payback" back then rather than "return on investment," because we knew we were not ready to convince the financial community that energy savings could be measured precisely in ROI terms.

AUDITING FOR PERFORMANCE CONTRACTING

Into this milieu, Scallop Thermal, a division of Royal Dutch Shell, introduced the concept of guaranteed savings in North America. The Scallop approach was to guarantee a 10 percent cut in the utility bill, and then cost-effectively meet the customer's needs for less than the guaranteed 90 percent.

Such guarantees required that the energy service companies' auditing expertise become more precise. Through one painful experience layered on another, ESCOs learned what worked—and what didn't! Still the focus remained on the measures themselves and seldom extended to the broad operational implications the measures might effect; nor did we really consider how conditions might change over the life of the equipment, or the project.

As *shared savings* came on the US scene in the late 1970s, ESCOs rather loosely calculated a customer/ESCO split of whatever cost savings had been achieved each month. By the late 1980s, *guaranteed savings* was the dominant ESCO approach and it demanded more solid energy saving projections. Gradually, a protocol for measuring and verifying energy savings, as discussed in Chapter 5, became accepted.

In a fairly short time frame, conditions demanded that engineers be more accountable to the ESCOs, who were betting large sums of money on projected costs and predicted energy savings. It is a challenge that we still struggle with today.

The traditional energy audit, which typically assumed all current conditions would remain static for the life of the recommended measure(s), was used to rank measures by payback and served to prioritize equipment purchases. The focus was on the measures, not the facility or process. Dedicated engineers used name plate data, run times, manufacturers' payback claims to calculate their projections. They fit the formula.

BEYOND PAYBACK

As energy prices climbed in the 1970s and 1980s, ways to reduce energy consumption proliferated. Not all measures were successful in reducing consumption and too many had a debilitating effect on productivity. Skepticism and uncertainty began to cloud the energy conservation movement.

If a measure saves energy, but costs the owner more in lost productivity, engineers and ESCOs have not offered clients truly "cost-effective" solutions. And the owner has bought a "pig-in-a-poke." We were made aware that an audit that interferes with sales in a Hong Kong retail shop, for example, could cost more in lost revenue than the audit's projected savings. Similarly, lost revenues in downtime to implement a measure could erase all the potential savings values.

In our zeal to save energy, we have too often narrowed our focus to immediate energy saving techniques. When saving energy was a whole new concept, the struggle to identify energy conservation opportunities, determine what equipment to install and calculate projected savings was almost overwhelming. Now that particular struggle is pretty much behind us, we need to take a broader perspective.

Owners do not buy "energy," they buy what it can do for them. To save energy effectively, it's essential to look at energy as a component of the total operation. It is essential to recognize that energy permeates every corner of a facility and every part of a process.

CONSERVATION VS. EFFICIENCY

CONSERVATION EVANGELIST

It is a mark of how little we have progressed that so many still use "energy conservation" and "energy efficiency" interchangeably. *Conservation* means conserving, using less. *Efficiency* means using what must be used as efficiently as possible. It is quite possible to do a quality audit; and, in the process of making a process more productive, actually recommend the amount of energy used in a facility be increased. As an example, Duke Solutions (now part of Ameresco) changed a two-pass gas curing process in a carpet factory to a one-pass infrared process. Energy consumption and costs went up, but the amount of energy per unit of product went down. That is efficiency, but not conservation.

Conservation to some still means deprivation and doing without. The word conjures up visions of President Carter sitting by the fireplace bundled up in his sweater declaring that the 1970's energy crisis was, "...the moral equivalent of war." This, in turn, reminds many of the old Emergency Energy Conservation Act that set Emergency Building Temperature Restrictions (EBTR), which had our children trying to write with mittens on their hands, all bundled up in cold, dark classrooms. American ingenuity hit new highs as occupants figured ways to fool the thermostat.

Our tunnel vision, surrounded by the conservation ethic, led to graphs, such as the one shown in Figure 2 comparing energy costs to personnel costs, which was actually published by the U.S. Department of Energy!

Such a graph was generally accompanied with an expression of concern—often approaching a diatribe—about energy conservation causing absenteeism and/or lack of productivity. Further, there was usually something about oper-

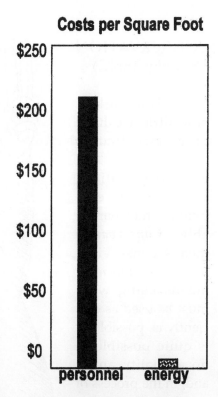

Figure 1-2. Personnel vs.
Energy Cost

ating costs at $200/sf. And energy was $3-4/sf.—and saving $2 in energy costs only to lose $30 in personnel losses was not good business. Most of us were indignant at the time and pointed out the absence of any direct relationship between the two factors.

But just maybe there was something to it. Those, who fuss about energy conservation and compare the value of energy cost savings to the losses in personnel time and productivity are justified if the auditor's conservation blinders kept him/her from looking beyond the equipment itself. Energy *conservation*, taken in isolation without regard to other conditions, might have caused illness and lost productivity. Energy *efficiency*, which considers the work environment and how energy is used throughout a facility, would not. Even today, conservation typically considers energy apart from other concerns, while efficiency considerations place energy smack in the bigger work environment picture.

In the 1990s, many of us became really irritated with the numerous magazine articles that continually explained that our indoor air quality (IAQ) problems were the result of the "energy efficient building." Again, an audit, focused on *efficient* use of the energy that was needed for an effective work environment, would not recommend measures that would have a deleterious effect on the indoor air quality. An audit only looking to conserve energy, however, might have such a result. Unfortunately, the writers of those articles never figured out the difference between an energy *efficient* building and a building focusing on conserving energy. And to our greater misfortune, neither did many auditors.

[*In a quick* mea culpa, *Shirley takes the blame for much of this problem. As part of the earlier referenced Saving Schoolhouse Energy project, which she directed, the federal government also charged the project with an evaluation of the Public School Energy Conservation Service program. Drawing from this program title, she used the term,* energy conservation measure, *to describe the specific energy saving recommendations. From this use, the "ECM" took root. It may not have changed a thing, but she now wishes, rather fervently, that she had used the term,* energy efficiency measure.]

Thus, through the years, by using the terms, *conservation* and *efficiency*, interchangeably we have all too often imposed on ourselves limits in what an energy audit could be, and should be. In doing so, we have left the door open for those who raised IAQ and lost productivity worries, which seem to come back to haunt us on a regular basis.

THE AUDITOR

As the complexities of the audit have increased, the demands on today's auditor may, at times, seem insurmountable. We are very mindful that we are now adding a far broader perspective to facility/process energy analyses. It is, therefore, important that we should not lose sight of the fact that really good auditors have been gradually incorporating many of these broader concerns into their work. It has, however, been on a rather ad hoc basis. The time has arrived to *systematically* weigh these issues with respect to each client's needs and incorporate them into an *investment grade audit*.

Engineers are comfortable with science, but "art" is another matter. An audit that meets all the needs touched on in the next chapter, and throughout the book, will not happen overnight. Delivering an Investment Grade Audit (IGA) is a professional growth process. And for every client, the demands will vary. An IGA auditor must venture into unknown waters that go well beyond the formulas in an engineering handbook. An IGA requires subjective judgment of people factors and a way to systematically determine the impact certain people factors will have on the paybacks. Somehow the risks inherent in implementing a measure in a certain facility or in a particular process, given its specific conditions, must be identified, the management/mitigation strategies determined, and the real cost of the measure/project calculated.

Not surprisingly, engineers, who can perform an IGA, are in short supply and in increasing demand. ESCOs are always searching for auditors, who can perform a quality IGA, thereby reducing an ESCO's risks. In fact, because ESCOs cannot afford to have their valuable engineers performing audits that do not lead to a project, ESCOs typically charge owners a premium for an investment grade audit if the project does not go forward.

Demand, in time, creates supply. Engineers, who have gradually modified audits to reflect the human element and operational concerns are becoming more deliberate in their efforts to do so. More engineering firms are incorporating risk assessment procedures in their auditing protocol. The owner, however, has to be leery of the engineering firm or ESCO that claims it is offering an IGA, but continues to do business as usual.

THE OWNER'S EXPECTATIONS

Owners have a right to demand that an energy audit be a sound guide to investment, whether guarantees are involved or not. As those, who want the quality of a performance contract without the ESCO guarantees and fees, begin to appreciate what an IGA can offer, the demand for a higher caliber of audit will grow even more. The IGA is destined to become increasingly popular among more knowledgeable energy efficient consumers, but it will take some time before it is a standard in the industry.

Owners, who want this quality audit to guide their investments today, have difficulty finding those capable of performing *real* IGAs outside the ESCO industry—and all too often within it. Historically, the auditor's involvement ended with the delivery of a report to the owner and the collection of a fee. The engineer's fee did not carry any assurance of predictive consistency. There was no immediate accountability.

With an IGA, the owner should insist upon, *and get*, cost and energy savings projections which the auditor will stand behind. Owners who ask for the level of accountability associated with an IGA, are more apt to find it. An owner, seeking a quality IGA, can screen analysts by comparing the auditor's predictions for previous projects with the results actually achieved. An auditor, who consistently gets results within the 90-110 percent range of original savings predictions, is an engineer who has mastered the art.

When one finds auditors ready to stand by their cost and savings projections, owners (as well as ESCOs) should expect the audit to take slightly longer and to cost more. In addition, the owner should recognize the need to make management and staff available to the auditors as they assess the people risks associated with a given facility or process and the measures being contemplated.

When auditors become more sensitive to all of an owners needs, an IGA report is apt to lead to a Master Plan that addresses other concerns, such as indoor air quality, operations and maintenance, commissioning, emissions reductions, etc. The value of a Master Plan to the facility owner is discussed in Chapter 10.

Technically, we knew we have been on increasingly solid ground with our calculations as to what individual "energy conservation measures" could do. We have evolved better ways to measure and verify savings so we can assure clients our predictions have been realized. Yet, the bar graph from US DOE, shown in Figure 1-2, comparing energy costs to personnel costs and the growing fervor in the 1990s about energy efficient buildings creating sick building syndrome kept bumping up against our efforts. The concerns were there—coming at us from all sides—we just failed to see that the self-imposed limits of the traditional energy audit would not serve the best interests of our clients.

The call for an *investment grade* energy audit is becoming louder. Financiers are increasingly expecting it. The ESCOs are increasingly demanding it. The motivation for so many to change the name of their traditional energy audits to "investment grade energy audits" is fairly obvious. But, as it is stressed throughout this book, changing the name is not enough. The challenge to the audit recipient is to sort through the verbiage and determine if the needed investment guidance and the elements of an IGA have truly been incorporated.

Ironically, if we get really good at the money side of energy, we may just have to satisfy those CPA-type demands called for by the feds in 1977 after all. How painful it is to admit the feds were at least partially right after all our snide remarks back in the late 70s.

When all is said and done, a quality energy audit must stand up to the careful scrutiny of owners. But the scrutiny of bankers and other investors will be even more demanding. An IGA is at the heart of a "bankable project." Hence the term, *INVESTMENT GRADE* energy audit.

Any thing less no longer adequately serves the owner, the contractor, or the investor. This book is about "raising the bar" in our auditing procedures to meet this growing need.

References
1-1   Federal Register Vol. 42, No. 25, June 29 1977.

1-2   Choksi, Nick, "Energy Audits for Health Care Facilities," *Energy Engineering Technology: Proceedings of the First world Energy Engineering Congress*. 1978. p. 79.

1-3   Readers, who wish information on basic technical auditing techniques, are referred to the *Handbook of Energy Audits, Sixth Edition* by Albert Thumann. Published by The Fairmont Press.

1-3   The previously cited 1977 Federal Regulations established 3 audit classifications. Class C were audits performed by owners with the guidance of a workbook.

1-4   Albert Thumann, *Handbook of Energy Audits*. The Fairmont Press Georgia. 1983. pp. 3-4.

Chapter 2

Why the *Traditional* Audit is Just Not Good Enough

MAGINE YOURSELF walking into an investment broker's office and being greeted with, "What can we do for you today?" After expressing your desire to invest in something with a strong return on investment, the broker says, "We have just what you are looking for. Please understand that some of our predictions are off by as much as 400 percent, but due to our strong technology base we are happy to say that our average error is just under 25 percent."

Remember the duck, who ran around in insurance commercials yelling "AFLAC" even though no one listened? After several commercials, there was one where Yogi Berra says, "And they give you cash, which is just as good as money," and the poor duck was struck dumb. At this point in the broker's explanation, you probably feel a lot like that duck, but being naturally polite, you respond, "Please explain."

With a smile displaying ultimate confidence in the product (and probably figuring he's got a real live sucker before him) the broker says, "We are able to achieve 74.9 percent of our predictions because we base our work on the best technology and engineering principles available."

At that point, you cease wondering why someone whose job it is to make you wealthier is called a "broker."

Research has shown that until recently the 74.9 percentage figure was what the auditing community on average offered the customer with the traditional energy audit. The wake-up call came from Texas A&M when they evaluated the work of *pre-qualified* energy auditors five years after projects had been implemented.[2-1]

Their findings are shown in modified form below in Table 2-1.
When co-author and Texas engineer, Jim Brown, reviewed these results, he told his fellow engineers, "It's time to raise the bar."

Table 2-1. Energy Saving Retrofit Results:
Texas LoanSTAR Program

Annual Utility Cost	Estimated Cost Savings	Measured Cost Savings	Measured/ Estimated Cost Savings
$584,972	$411,066	$165,520	40.3%
$261,276	$161,956	$249,209	153.9%
$139,782	$118,179	$143,980	121.8%
$649,474	$373,621	$709,271	189.8%
$41,738	$10,485	$35,090	334.7%
$57,784	$26,367	$46,476	176.3%
$84,482	$20,400	$30,496	149.5%
$1,111,240	$303,435	$95,522	31.5%
$105,296	$42,049	$90,034	214.1%
$114,741	$41,235	$59,573	144.5%
$188,269	$44,881	$177,447	395.4%
$54,048	$5,768	$4,404	76.4%
$278,985	$65,955	$83,333	126.3%
$64,197	$15,682	$69,189	**441.2%**
$147,743	$83,960	$33,991	40.5%
$239,569	$109,334	$48,903	44.7%
$3,677,292	$664,589	$82,416	12.4%
$52,342	$33,094	$43,339	131.0%
$76,440	$12,754	$38,281	300.1%
$55,064	$17,240	$7,691	44.6%
$47,229	$27,069	$19,660	72.6%
$171,504	$83,416	$31,649	37.9%
$77,532	$81,077	$25,998	32.1%
$473,094	$330,984	$18,283	**5.5%**
$364,754	$128,525	$96,240	**74.9%**

For a time, when we did not get the expected results, we blamed the manufacturers' efficiency ratings. We even said that the problem was that manufacturers were testing their equipment in a laboratory setting and not in "real life" situations. And we still didn't get it.

It's time to let "real life" in the energy auditing door. Before going any further, let us hasten to add that a growing number of engineers have been gradually raising the bar. They have moved beyond the formulas in the engineering handbooks and are now drawing on their experience to fine tune their predictions.

LIFE AIN'T WHAT IT USED TO BE

For several decades the traditional energy audit, so familiar to engineers, served the energy efficiency industry quite well. Energy management plans were prepared, contracts for efficiency projects were written and work went ahead. Energy was saved and most often, clients and contractors were reasonably satisfied.

Energy managers used audits to plan energy improvements within their facilities and to justify proposed funding for those improvements. Energy service companies (ESCOs) used the audits to develop projects for clients. Because of the limitations of the basic audit, "risk cushions" were written into contracts to account for the uncertainties that seemed inevitable. As a caution, ESCOs seldom guaranteed more than 80% of the savings as predicted in the audit calculations. In fact, the guarantee was very often in the 60—70% range, which meant that many of the potential investment dollars (and potential project benefits) were left on the table. Dollars that could have bought more equipment and more savings.

It became obvious the goals of energy efficiency and fiscal responsibility were being "short changed." Notably, it was the introduction of financial accountability with its uncompromising drive toward project profitability that forced recognition of the "elephant in the room." Neither the owners nor the energy auditors were ignorant of the risk avoidance techniques that had been developed. But, until financial accuracy became as important as technical accuracy, no one was going to acknowledge the presence of the elephant.

It was concluded that if audits could come closer to the prediction of actual savings over the life of a contract, projects could be larger, better for the client, more satisfying to the engineer, and more profitable for the ESCO. Consistency of audit predictions and results would also provide an increased level of confidence to financial houses asked to provide funding for energy efficiency improvement.

The *traditional* audit was not wrong; the type of information historically gathered in a survey is as important today as it ever was. Basic to understanding of an investment grade audit (IGA) is the realization that the traditional audit must serve as a starting point.

The foremost limitation of the traditional energy audit has been its allegiance to the "snap shot" approach. It's like someone took a picture and all conditions were frozen in time. Auditors assumed conditions observed during the audit would remain the same for the life of the equipment or the project.

Such fallacies led us to assume the condition of related equipment on audit day as well as the current hours of operation and use of space would remain the same. We had a good picture of the historical situation and of *current* conditions. All this was critical information and is still needed as a starting point, but it is not enough. An

THINGS CAN CHANGE
AFTER THE SNAPSHOT IS TAKEN

audit today must support a multi-year energy efficiency project and provide quality investment guidance. If performance contracting is involved, the audit must predict future savings from measures taken by the ESCO. If Master Planning is involved, those predictions must evaluate future savings to be gained from *future* installations/modifications. Master Planning will be treated in more detail in Chapter 10.

The traditional audit too often also limits itself to energy *conservation* measures, not necessarily meeting the owner's energy *efficiency* needs. Today those needs also require attention to unique process requirements, supply availability, energy security and information systems.

Actually the term "traditional" is something of a misnomer for the audit that has been in use for so long. It brings to mind Tevye's song, "Tradition," in *Fiddler on the Roof*, where he reminded us of the values of doing things the way that they have always been done. We recognize that the "traditional" energy audit does not have this sense of sameness. Since energy audits began to be heavily used in the late 1970s, they have been upgraded as we learned more about equipment operations and efficiency characteristics. Technologically, audits have been in a state of constant evolution with more, and generally better, data gathered as engineers learned to interpret what they observed in terms of energy efficiency possibilities.

This approach has served us very well, but with all the improvements, it has remained basically a technical study of conditions as they exist at the time of the audit; hard data to be sure, but the analysis was limited to the benefits of implementing specific measures encapsulated in existing conditions.

As the energy efficiency industry matured, projects became more complex and risk analysis became more sophisticated, it became obvious that the need for information had grown beyond what regular audits produced. At the same time, energy efficiency project financing was getting easier to find, but financiers were asking more sophisticated questions about risk and basing interest rates more closely on the answers.

The assumption that all current conditions would stay the same was the norm. But buildings and systems are seldom, if ever, static. They are typically dynamic with changing functions. What's more, they are populated and operated by people who simply will not behave in predictable, consistent ways.

What is needed today is an *investment grade audit* that combines the traditional with a forward looking risk assessment component; a step beyond technical fact-gathering to a consideration of what is likely to happen over time. The more sophisticated IGA approach looks at the entire operation and the owner's needs, embracing energy concerns and operational matters into the future.

For many engineers, dedicated to the certainty of hard data, whose careful calculations have been based on tried and true formulas, this forward thinking remains a new and, in some cases, a frightening departure. It requires SUBJECTIVE JUDGMENT—and all the horrors that concept suggests to the engineer. An engineer, who has traditionally relied on his or her trusty handbook, shudders at the idea. For those who grasped the concept and moved ahead, however, the results (in client satisfaction and expanded business opportunities) have been well worth the work and fortitude required to master this new approach.

Some energy engineers, particularly those working for ESCOs, have adopted a true investment grade audit mentality. These engineers now feel they should be able to regularly accomplish savings results in the range of 90 to 110 percent of what their audit predicted. For those who offered guaranteed results, gone are the overly large risk cushions, which limited project size and profit potential. As an important additional benefit, financial houses tend to view an investment grade audit as added "due diligence," helping to lower the cost of project financing.

PUTTING THE *TRADITIONAL* AUDIT TO WORK

A well done, well-documented, traditional audit is an integral component of the IGA... a vital component. The information gathered provides the basis for the questions that must be asked to extend the "snapshot" to a long-term estimate of the conditions that will assure the success of an energy efficiency effort *over the life of the equipment and/or project.*

The content of a solid audit has been refined over the years to a rather standard, and useful, format that can provide comparisons from one project to another. It has proven especially valuable when considering projects in similar facilities. More importantly, it offers a format that can be augmented to incorporate the risk components and mitigating strategies essential to an investment grade audit.

It is worth repeating that procedures used in the traditional audit have not been wrong; they simply have not gone far enough. "Paybacks" have given us false courage. Somehow, over the years, we have used paybacks almost exclusively as a means of ranking energy conservation measure to be sure we implemented the most cost-effective opportunities. This has, of course, been very helpful, but in today's world, it simply no longer meets our needs.

If energy efficiency is to be viewed as an investment, *and it certainly should be*, then the payback figure must be upgraded to serve as a reliable guide for the investors who wish to finance energy efficiency projects. The energy audit report must be more than a guide to save energy, it must also give direction for more effectively and efficiently managing a process and/or facility's investment portfolio.

It is also worth restating that the reader is expected to already have a basic understanding of the traditional energy auditing procedures, for this book builds on that knowledge.

Historically, we have skirted the implications of the human element in energy auditing. "Paybacks" have been assigned to certain measures in multiple applications when we knew full well they would not perform in exactly the same manner under differing conditions. As early as 1983, an evaluation of the first eight cycles of the U.S. Department of Energy's Institutional Conservation Program by The Synectics Group, Inc. revealed that up to 80 percent of the savings in an effective energy management program could be attributed to the energy efficient practices of the operations and maintenance (O&M) personnel.[2-1] In other words, as little as 20 percent of the savings could be attributed to the actual hardware, but we continued to make calculations as if a piece of hardware was always going to operate in the same fashion—and save the same amount of energy—under vastly different conditions. The values assigned to various types of measures from US DOE, as depicted in Figure 1-1, lent additional credence to this belief.

When asked, astute engineers will tell you that they do not assign the same paybacks to identical measures, even in very similar facilities. But when asked to explain how they arrive at the calculations for this difference, most shrug their shoulders and mumble something about "just knowing… " It is time we take that "knowing" and systematically apply it to our auditing protocol.

To repeat: it is past time for those of us who wish to predict savings with any degree of confidence to turn that "just knowing… " into a systematic approach. It is time to give owners, ESCOs and financiers an *investment grade audit*.

For every significant element in a traditional audit, we now must take a hard look at the related conditions and their impact on the total operation. Certain questions must be posed for *each* recommendation:

1.   How will this measure affect… ?

2.   What are the risks if this measure is implemented? And, what are the risks if it is not?

3.   What are the risks inherent in predicting a certain level of savings over time?

4.   What risks can be mitigated and what will the mitigation strategies cost?

5.   When the costs of the risks, which can't be mitigated, and costs of the accepted mitigating strategies are factored in what is the new payback?

6.   Finally, when all the above is considered is the measure (or the project) still worth doing?

Underlying this catechism is an assessment of the "people factor" and related operational concerns.

To get the answers, the auditor faces new challenges in knowing the right questions to ask, what to observe, and how to analyze these findings.

Then, comes the hardest question: how do we know if we are doing it right? The ultimate determinant is in the results. How close do the actual savings match the predictions? Fortunately, there is a yardstick for measuring the quality of an IGA—predictive consistency.

PREDICTIVE CONSISTENCY

The goal of the IGA auditor is to reduce the level of uncertainty that surrounds the implementation and maintenance of energy efficiency measures as well as predicted potential savings over time. A measure of an IGA auditor's success is *predictive consistency*. How often does the auditor come within ±10% of his/her predictions?

Obviously, some degree of understanding should be accorded those firms, who still venture into the unknown, attacking unusual problems requiring untraditional techniques. But for most owners, faced with a plethora of engineering firms and ESCOs, who claim they offer IGAs, the consistency of the predictions must be assessed. It is the only way to weed out those who have changed the name, but not the game.

Predictive consistency can be tested by determining how consistent an auditor normally is in predicting the level of energy savings that will be achieved *over time*. It is not enough to be right for the first six months, or the first year; nor is it sufficient to get just one project right. The savings have got to be there for every year of the calculated payback period. For every year of a project.

For meeting the self-funding promises. And for project after project. *Consistently!*

It is easy to over play the "consistency" argument at this point. We want to hasten to say that the best of firms will have a problem on occasion. The two cases offered by Jim Brown on the following page highlight the fact that no firm is infallible.

At this point, the audit goes beyond science and becomes an art. The human factor must not only be assessed, but it must be paired with potential energy measures to ascertain what impact occupants, management, maintenance and operational behavior, etc. will have on those measures. For example, in facilities where confidence in the human factor receives a relatively low score, measures that are practically people impervious, such as insulation, can be looked on more favorably. While measures, such as controls—particularly if overrides are readily accessible—carry a greater risk of human intervention. In every case, the payback and predicted savings results must be adjusted accordingly.

In such circumstances, possible mitigating strategies and their costs must be carefully considered. Thus the consideration of risk enters the IGA equation. As do the costs associated with mitigating those risks.

All of this information is assembled into a report, which may form the basis for a contract. Often this audit is referenced as a part of a final contract. When legal entanglements are on the horizon, it pays to be right.

A quick definition of the risks that are assessed in an IGA can probably be captured in a short question, "What are the factors that might prevent the achievement of projected savings?" The auditor must look at people, management attitudes and policies, overall facility considerations, O&M capabilities/practices, prospects for change in functions, energy price outlook and any other factors that will impact a project over time. An IGA auditor considers many "What if… ?" questions. Most of the answers require evaluative judgments that were seldom taught in engineering school.

No One Is Infallible: Two Cases in Point

No engineer or energy auditing firm, which has been in the business for any length of time, can point to a perfect record. Our firm is no exception. We have, on at least two occasions, failed miserably in our attempts to produce the predicted product.

From the first, we learned that time restrictions cannot be allowed to pressure us into incomplete measurement procedures. You've run into the same scenario: The request for professional services was delivered in February. Those services were to produce the systems analysis and calculations needed to justify the replacement of two chillers that were well beyond their twentieth birthday. The school board, however, didn't get around to voting on the proposal until May. Their decision was to replace only one chiller and the selection of which one to replace was left to our "better judgment." Since the sizes of the chillers were large, they had a shipping cycle of eight to twelve weeks and we had all of four days to measure operation levels, check old maintenance reports, and talk to the maintenance men responsible for the system operation before making our selection. Since classes were still going, we did not have the option of pulling the heads and checking the internal components for wear. We finally made our decision using grossly inadequate data. We ordered the chiller, got it delivered and finished the installation just two days before classes started up again that August. The old chiller that had not been removed had been sitting quietly by all summer long awaiting the installation of the new unit, then, when we hit the switch to restart the system, the old boy just refused to come back to life! Maybe the other chiller would have died that summer as well, but you couldn't convince the board members of that! We were the experts and we should have known better!

The second taught us a lesson in "people predictability." It was an old school with sorely inadequate and antiquated equipment... it was a building on its way to shutdown with little money left in the school budget to do anything but watch the inevitable demise. There would be energy savings, but unless the school administration was willing to accept some stringent operating restrictions, the savings would not come fast enough to satisfy the lender. So, conditions were communicated and promises were pledged.

The money was invested... the new equipment installed... then, the school board changed members and the superintendent moved on. The first thing the new board did was rescind the restrictions. The first thing the new superintendent did was tell the world that predicted savings weren't being obtained. You know the rest of the story.

Jim Brown
President, Energy Systems Associates

FACTORS TO CONSIDER

The people factor in an energy efficiency project can never be overlooked and this consideration starts with **top management** of the target facility. Too often, top management exists in its own rarefied environment as depicted on the next page.

Seeing issues from management's perspective and getting on their agenda is of strategic importance. If top management is committed to improvement of energy efficiency and has a clear policy toward that end, it is a big step toward a successful project. One very successful engineering firm will not undertake a detailed IGA unless the client has a written energy policy in place, or is willing to begin the program with the writing of this policy—a policy that will be clearly transmitted from the top down through every level of the organization.

Throughout the **organization** attitude, interest and understanding count. An IGA will look at the general behavior of the occupants of the facility in question which actually reflects management's visible commitment. This is especially true when **operations and maintenance (O&M)** are considered. If the head of maintenance and operations feels his domain is being unfairly invaded, or if he or she feels an efficiency project reflects poor performance on their part, the project is almost sure to fail. We should never lose sight of that fact that the very word, *audit*, connotes an accounting of past performance, which, all too frequently, has been regarded by O&M personnel as a threat.

Also at the maintenance and operations level, the IGA will include an assessment of the **needed skills and training of personnel**, who will have to operate and/or maintain energy systems within the facility. This assessment will also include a consideration of worker turnover (which has a major bearing on training needs). A look at the way current **maintenance scheduling** and how it is carried out will give an idea of how well systems may be maintained over time. Does there appear to be a systematic approach to what should be routine maintenance or is the work "breakdown oriented"? Are maintenance records kept and preventive maintenance performed as scheduled? **Housekeeping** is

also an indicator of future performance because it reflects allocated budgets, attitude and/or management interest. All of these people factors will make a difference in the way a project performs over time and must be assigned a value when calculating the risk level in a project plan.

In considering the dynamics of function and use over time, **facilities** are a key factor to consider. In the early 1980s, the introduction of energy consuming computer labs caused complications in a number of energy performance contracts in schools... contracts written before computers became ubiquitous. In too many cases, the contracts were based on an audit that said, in effect, "This is a school and schools are pretty much alike and will stay that way ...forever." Too often, ESCO and customer were left to negotiate their way out of difficult problems, since change had not been anticipated and accounted for in the audit—or the contract.

Process changes occur. Industrial facilities eliminate or add product lines or functions that change energy demands. They also may add a work shift or cut back on hours. Determining the likelihood of such occurrences is not a precise science, but recognizing the possibilities allows for contingencies to be addressed in the audit report; so the matter can be handled in manner that does not disrupt, or destroy, the project.

Sometimes we must state the obvious to the clients and investors. The owner and lender need to recognize that increased usage increases consumption. And just as obvious, an ESCO should recognize that decreased usage reduces consumption. Both possibilities should be recognized by all parties before the project begins.

The more that is learned about such possibilities; the lower the risk of future problems. Since money must be set aside to deal with the risks, reducing those risks means more of the investment can be used to improve the facilities/processes, the savings and the profits.

Measurement and Verification

Measurement and savings verification (M&V) and the IGA go hand in hand. M&V requires an established and clearly defined historical baseyear as well as the necessary procedures to create an annually adjusted baseline. Since the auditor can't calculate potential savings or cost-effectiveness without these reference points and they are integral to an M&V procedure, the auditor must be cognizant of the M&V implications.

The extent to which the owner intends to use M&V can also affect the audit. If a baseline is too dynamic to establish a clear reference point, the measure may not be recommended. Or, it may be implemented with stipulated conditions and typically, outside a guarantee package. If M&V for a certain measure is too costly, the measure may be removed from consideration. The relationship between M&V and the IGA are explored in Chapter 4.

EXPANDING CONSIDERATIONS

As investment grade auditors become more sophisticated and sensitive to the issues surrounding the recommended energy efficiency measures, they will come to realize that there are more opportunities to move up the value chain—and more ways to serve the client more effectively. Illustrative of this opportunity are issues related to energy information and energy security. They are discussed below to remind us that there are already many more concerns around the next corner.

Energy Information
In the energy sector, whether referring to engineering consultants, ESCOs or their clients, it is safe to say that the majority do not:

a)   effectively use what they know;
b)   fully appreciate what they don't know; or
c)   have any plans to systematically obtain the needed information.

In 2002, *Strategic Planning for Energy and the Environment* launched a series of articles regarding "Information Technology Basics for Energy Managers."[2-2] These articles serve the critical need of understanding information technology (IT), its capabilities, tools, and lexicon. To make guidance truly practical, however, the energy community needs to assess the content that will flow through these IT conduits.

While getting a handle on all aspects of IT and IT's capabilities is incredibly valuable, we should not lose sight of the fact we are still talking TECHNOLOGY. It conjures up pictures of solid gold conduits with nothing flowing through them. It is reminiscent of early massive computer application days when OPEC installed its state-of-the-art mainframe data and analysis system. Since obtaining data was frustrated by the mistrust among the OPEC members, the folks at OPEC headquarters proceeded to regularly input data from the readily available information in Oil Daily.

Any energy IT system must be measured by its usefulness. And its usefulness is dependent on the quality of the information flowing through its veins. Fundamental to effective energy management planning and its integral information system is data obtained through a quality audit. The rather sterile data available from an Oil Daily equivalent, the traditional audit, lacks the richness and usefulness of what an IGA can deliver.

Energy Security

Only when the lights go out, the motors stop running, and the air becomes stifling does the average person realize what a critical component energy is to our comfort, our ways of doing business and our total life style.

At the beginning of this decade, the drought in the Pacific Northwest, the mismanaged mess in California as the state attempted to deregulate, and grid problems around the US gave us a taste of what it's like to do without energy. It has become increasingly obvious that energy is our economic lifeblood. Since the terrorists brought their war to our shores, we have become increasingly aware of how incredibly vulnerable our country's energy infrastructure is. Secretary Ridge has noted on more than one occasion that our energy infrastructure is very vulnerable.

Energy managers and the auditors that serve them are only beginning to realize that energy availability is a critical component in energy planning. Energy may be a small percentage of an operating budget, but it is vital to the work environment and keeping processes operating.

Taken in this context, availability could become far more important than price. Energy makes up less than one percent of the total budget for Johnson & Johnson, but the absence of energy could wreak havoc with the company's operations—*and have a far greater impact on its budget.*

While surveying energy use in a facility or a process, a top-notch auditor should assess and offer guidance on energy security matters. The audit report should help the manager identify critical energy needs, help prioritize vital demands for limited energy resources, and determine the availability of alternative energy sources to meet those needs.

These requirements, which go far beyond an assessment of energy efficiency opportunities, recognizes energy as a critical component of the total operation. It compels the auditor to identify ways for his/her client to have the energy resources necessary to operate under a range of scenarios. We are now going beyond the usual stand-by generation and looking at distributed energy resources, including combined heat and power as well as renewables.

Simply put, the auditor, in this day and age, has not fulfilled his/her responsibilities to the client if availability considerations are not addressed in the audit report.

As the energy efficiency industry has matured, it has become increasingly obvious that the traditional audit is no longer good enough. It probably helps explain why so many firms have rushed to change the name of their audit.

As projects have grown in size and complexity, there has been a growing recognition of the need for something beyond a snapshot on which to base decisions. A much more comprehensive approach has become necessary. ESCOs have helped to push this envelope as it became clear that the ability to guarantee results at a level of 60 to 80 percent of predicted success was not good enough. Bringing together a quality technical assessment (traditional audit) with a much more subjective assessment of human factors and longer-term facility considerations is allowing the industry to grow and gain credibility with customers and with the financial community. The engineer who can produce a solid, investment grade audit makes a major contribution. It's a combination of science and art... and it works.

References

2-1   Jeff S. Haberl, et al. *An Evaluation of Energy-Saving Retrofits from the Texas LoanSTAR Program.* Table 5. Work sponsored by U.S. Environmental Protection Agency. 1996.

2-2   Capehart, Barney and Paul Allen, Klaus Pawlik & David Green. "Information Technology Basics for Energy managers—How a Web-based Energy Information System Works." *Strategic Planning for Energy and the Environment.* First article Vol. 22, No. 3 Winter 2003. pp. 7-24.

Chapter 3

Weighing Human Behavior

VEN THE BEST of engineers can go only so far in assessing energy saving opportunities with technical calculations. When it comes to performing an audit, it would be a lot simpler if the buildings remained empty and processes were all performed by robots. Nice predictable robots.

Shirley's favorite school custodian, Ziggy, always took great pride in getting the school ready for the new school year. As the school's principal, Shirley would tell Ziggy in August how great everything looked. Ziggy always responded, "Now if only we could keep it that way, but I suppose we have to let the teachers and the kids in." That's where we are in the audit process, we are just going to have to let the people in.

When we do that, engineers face their greatest nightmare… incorporating *people* into the predictions. People are incredibly unpredictable; so how can we factor their actions into our savings predictions? Dealing with nightmares and things that go bump in the night seem easier, but it's not quite as overwhelming as it seems at first glance.

Auditors have for years taken into account the impact people might have on the designs or changes they are considering. Equipment has been selected based on the operations and maintenance staff's ability to operate and maintain such equipment. Controls have been judged cost-effective in many facilities, because the people could not be counted on to routinely turn things off.

To move ahead, we have to take as a given that for savings predictions to stand the test of time, we must consider the people who manage the facility, the people who operate and maintain it, and the people who occupy it.

Every engineer has a locking-thermostat-cover story. From

the teacher who took off her shoe and busted off the cover, to those who have seen ice packs or Bunsen burners employed in the hopes of getting the ambient temperatures cool enough or warm enough to suit. Some "tamperers" have become very ingenious in their temperature control efforts.

In the late 1970s, a popular story was the one about disconnecting the thermostats and putting small motors in the plenum; so they would run when the thermostat was adjusted. In so doing, of course, the occupants thought they had control of their environment and were satisfied that they had "fixed it." With this slight of hand, they thought they had a more comfortable place to work.

WHERE THERE IS A WILL

JUDGING BEHAVIORAL IMPACT

The IGA rests on the auditor's ability to take a look at the people, at all levels, and judge what their impact might be on a contemplated measure. There are key signs that help us make that judgment. As we grow in our ability to make such evaluative judgments, the risks will drop, and the predicted level of savings will be more certain. The challenge before us is to consider the key variables related to these risks and systemizing the process.

The Investment Grade auditor is neither Freud nor Ann Landers; so accurately predicting *future ignorance* is not expected. A noted southern statesman once said, "If ignorance ever goes to $40 a barrel, I want drilling rights on that man's head!"

Getting the drilling rights is what the IGA is all about. To the client, risk is an accepted expense. If you reduce the client's risk, that organization can operate more effectively.

The beauty of all this is that money follows risk. Risk management can become a profit center in your operation. Success in the energy efficiency industry will belong to those who most effectively and systematically:

- identify the risks,

- evaluate them accurately,

- develop appropriate and cost-effective mitigating strategies, and

- put the right price tag on the risks they accept.

THE PEOPLE FACTOR

Since the people factor has been the weakest aspect of auditing procedures in the past and since these concerns must permeate all the issues raised in this book's subsequent chapters, we will lay some groundwork here. Later, people factors will be revisited as critical risks in Chapter 7 and in related potential mitigating strategies in Chapter 8.

Management Policies and Commitment

If an owner, or client wants a successful energy savings project, then they must be willing to make a viable and visible commitment toward the success of the program. The best way to develop and demonstrate this commitment is through the implementation of an Energy Management Policy. A Policy that ends up in a notebook on a shelf that no one ever sees will not suffice. A key to the success of an energy savings project is making sure that everyone in the organization is aware of the Policy and keenly aware of management's very strong commitment to implementing it.

An effective Energy Management Policy Statement addresses management's intentions, the savings goal, the authorization of an energy management position or department, actions to be taken, expected cooperation of departments and staff, evaluation procedures, and possibly some specific incentives. A sample Energy Management Policy from a school district in Texas is presented in Appendix B.

Some organizations include certain details in the Policy that might better be in an Energy Management Plan. A classic example of this is temperature settings for the heating and cooling seasons. The reason for this reaches back into the later 1970s and the federal Energy Building Temperature Restrictions. These EBTRs led people to believe that energy conservation meant freezing in the dark. A long painful history has taught us that reasonable temperature levels that are adhered to are much better than extreme ones that are not.

Look at the suggested heating and cooling season indoor temperatures shown on the following pages from Al Thumann's earlier *Handbook of Energy Audits*.[3-1] Sit back a moment and contemplate: can you guess the percentage of 105-pound elementary school teachers that will voluntarily set thermostats at 68°F in the winter? Or, how many CEOs will self impose 78°F in their offices on a hot summer day? Can you imagine yourself as the maintenance staffer who has to explain it to them?

Given these temperatures, it is not too surprising that a number of people in the 1980s raised the question about lost productivity in "energy efficient" buildings. For energy management

	A Dry Bulb °F occupied hours Maximum	B Dry Bulb °F unoccupied hours (set-back)
1. OFFICE BUILDINGS, RESIDENCIES, SCHOOLS		
Offices, school rooms, residential spaces	68°	55°
Corridors	62°	52°
Dead Storage Closets	50°	50°
Cafeterias	68°	50°
Mechanical Equipment Rooms	65°	50°
Occupied Storage Areas, Gymnasiums	55°	60°
Auditoriums	68°	50°
Computer Rooms	65°	As required
Lobbies	65°	60°
Doctor Off ices	68°	58°
Toilet Rooms	65°	55°
Garages	Do not heat	Do not heat
2. RETAIL STORES		
Department Stores	65°	55°
Supermarkets	60°	50°
Drug Stores	65°	55°
Meat Markets	60°	50°
Apparel (except dressing rms)	66°	55°
Jewelry, Hardware, etc.	65°	55°
Warehouses	55°	60°
Docks and platforms	Do not heat	Do not heat

	A 24 Hrs or less		B Greater than 24 Hrs
3. RELIGIOUS BUILDINGS			
Meeting Rooms	68°	560	50°
Halls of Worship	65°	560	50°
All other spaces	As noted for office buildings	50°	40°

Source: Guidelines For Saving Energy In Existing Buildings-Building Owners and Operators Manual, ECM-11

Figure 3-1. Suggested Heating Season Indoor Temperatures

I. COMMERCIAL BUILDINGS	Occupied Periods	
	Minimum Dry Bulb Temperature*	Relative Humidity
Offices	78°	55%
Corridors	Uncontrolled	Uncontrolled
Cafeterias	75°	55%
Auditoriums	78°	50%
Computer Rooms	75°	As needed
Lobbies	82°	60%
Doctor Offices	78°	55%
Toilet Rooms	80°	
Storage, Equipment Rooms	Uncontrolled	
Garages	Do Not Cool or Dehumidify.	

II. RETAIL STORES	Occupied Periods	
	Dry Bulb Temperature	Relative Humidity
Department Stores	80°	56%
Supermarkets	78°	55%
Drug Stores	80°	55%
Meat Markets	78°	55%
Apparel	80°	55%
Jewelry	80°	55%
Garages	Do Not Cool.	

*Except where terminal reheat systems are used. With terminal reheat systems the indoor space conditions should be maintained at lower levels to reduce the amount of reheat. If cooling energy Is not required to maintain temperatures, 74°F would be recommended instead of 78°F.

Source: Guidelines For Saving Energy In Existing Buildings-Building Owners and Operators Manual, ECM-1

Figure 3-2. Suggested Indoor Temperatures and Humidity Levels in the Cooling Season

policies and plans to be effective, they must take into account what management thinks constitutes a productive work environment. If the numbers don't reflect what management thinks, getting true management commitment will be extremely difficult. Without this commitment, there is no program.

When an Energy Management Policy is put in place, we are way ahead of the game. We have not only established the authority for operating the program, we have also created the means for implementation. This provides a clear road map to those who have been given program responsibilities. They now have a certain level of assurance that they will be fighting a "single front" battle... that is, they can spend their time creating and implementing an effective program without being quite so concerned that other staff members are sabotaging their efforts!

Occupant Behavior

Occupant behavior can range from willing compliance to outright rebellion. If the program isn't too restrictive and only mildly offensive to the building occupants, then hoped for results can be categorized as *agreeable acceptance*.

This acceptance will also be influenced by the manner in which the information is communicated. Effective communication will explain what is transpiring and why. It will also take into account the need for exceptions due to medical reasons or for unanticipated events. Preplanning of activities and timely requests for variances from the established norm will avoid most problems and/or discomfort.

Exceptions should, however, always be predicated on real need, and should not be allowed when the exception produces unacceptable expense, such as a peak demand spike, when a light sweater would have resolved the situation.

Operations and Maintenance

In terms of implementing an effective energy management program, the most important people in the buildings are the operations and maintenance personnel. The skills, manpower levels

and attitudes of operations and maintenance staff (O&M) are crucial considerations in determining whether or not certain measures will achieve the predicted savings.

This book might be taken as an "Ode to O&M," for we find
the need to support O&M efforts is frequently lacking, which is
quite an irony as O&M energy efficiency practices are typically the
most cost-effective measures that can be taken to reduce energy
consumption.

Somewhere in the ultimate business school that influences all
others, there is a plan that requires all budget-cutting schemes to
begin by attacking the maintenance department. Utilities bills are
going up... where do we get the money? Cut the O&M budget.
We need money for a new piece of equipment on the assembly
line? Cut maintenance. Facilities throughout the country are suffering the consequences of prolonged procrastination in the area
of equipment maintenance, O&M staff training and O&M manpower. As a result, equipment has deteriorated, along with O&M
staff morale.

An even greater irony is that deferred maintenance reduces
energy efficiency. A less efficient operation increases the utility
bill. How do we pay the higher utility bill? Cut maintenance. It's
a vicious cycle that goes nowhere but down.

An auditor, who has become sensitive to the people factor,
will include some level of analysis of past maintenance practices,
recent repair costs for problem equipment, preventive maintenance techniques, staff skills and attitudes. When the maintenance
department is not provided the training, tools or staff to keep up
with a facility's rapidly growing, and continually more technical
purchases, then problems arise. Not just equipment problems, but
attitude problems. These problems serve only to exacerbate the
situation.

This sensitivity inevitably leads to the conclusions that needs
within a maintenance department must be addressed before expensive new systems are installed. This concept has already become generally accepted by third party investment organizations,
but their remedy of choice is usually wrong. Their tendency has
been to charge more money to the project to cover additional

outside maintenance on the equipment, including the equipment which is being installed in the energy saving program. This leads to an unrealistic division of responsibility within the maintenance department regarding "whose job it is," and it separates the staff which is consistently on-site from the "ownership" of the program. Further, it breeds resentment among those who have put up with low level support for so long.

It would be of more value to the client, the maintenance department and the investors if the money were used to provide training and hire additional staff for the maintenance department. Of course, once in a while a project might require highly specialized technical support, but those projects are rare.

To summarize, the key, people-based issues that have been creeping into the traditional audit, but have become a must for an IGA are:

- management's real commitment to energy efficiency

- the resulting behavior of occupants based upon management's visible commitment

- possible impact a measure will have on facility/process; plant engineers' and facility managers' reactions and productivity implications

- operations and maintenance manpower, skills and training needs

- health and safety issues germane to the measures under consideration

- equipment constraints due to O&M limitations

- condition of energy-related mechanical and electrical equipment; the past O&M practices revealed by the conditions observed

- budget allocations for repairs and replacement

- attitude of O&M to energy program

•   the existence of or the potential for efficiency incentives.

People: A Source of Valuable Information
An incredible source of information and wisdom to help plan the needed changes in a facility or process is frequently overlooked: the people who have the day-to-day responsibility for maintaining and operating the facility and the people, who occupy that facility, are frequently overlooked. Given even the slightest opportunity, they will happily tell you what they would like to see changed. Much of this information can be incredibly valuable. It is also surprising how seldom these are grandiose schemes, but rather common sense remedies.

The added, and maybe greater, benefit is that these people truly appreciate being asked. It is always well to remember in conducting an audit that management is typically facility blind. They only open their eyes to truly see what is around them when something goes wrong... or they want their office fixed up.

The O&M rank and file are largely ignored. In fact, the better job they do in stretching their budgets, the more invisible they become.

This chapter has leaned toward the negative side of the "people factor" and has not sufficiently emphasized the positive matters you need to consider in auditing practices. In fact, for the building occupants, and especially the O&M staff, the word "audit" prompts negative connotations; specters of the IRS and fault finding seem to be floating around the auditor's head.

Experience tells us that looking into existing maintenance programs with an eye toward reinforcing their efforts will pay vastly higher dividends. Encourage them to talk to you and with you. The end result will be a more highly skilled staff, which is motivated to see the program work. If the Maintenance Department is working with you, risks are lowered and there is a much greater chance for success.

3-1   Thumann, Albert, *Handbook of Energy Audits, Second Edition, 1983.*
The Fairmont Press, pp. 128-9.

Chapter 4

Building
The M&V Foundation

EASUREMENT AND VERIFICATION of savings (M&V) is an essential part of any effective energy management program. Without measurement, there is no way to tell if your procedures are effective, if the energy efficiency measures are delivering as predicted, or if new procedures are warranted. Financial institutions, which have learned effective due diligence for energy efficiency projects, look for quality M&V procedures.

Further, the savings "yardstick" offers consensus benchmarks for industrial sectors and a means of evaluating engineers' predictions or ESCO's programs.

The best measuring instrument in the world, however, will not help if you do not have a reliable starting point. An investment grade audit (IGA) is the foundation of a solid measurement and verification effort. A major component of effective auditing is the establishment of baseline consumption and the identification of the conditions that caused that consumption.

A quick look at the essential elements of a M&V Plan will reveal how inextricably linked the two efforts are. M&V planning typically starts with the identification of the IGA recommended measures that will be implemented. This is usually followed by the following information.

- Definition of the boundaries of the savings determination;

- Documentation of baseyear[4-1] consumption, *conditions that caused that consumption,* and the resulting baseyear energy data, which will serve as a reference point to make annual baseline adjustments and calculate the savings;

- Indications of any planned changes that would affect baseyear conditions;

- Identification of the post-retrofit period during which results are expected and measurements will be taken;

- A clear indication of the set of conditions to which all energy measurements will be adjusted; i.e., how the adjustments will be made so current conditions are reflected in the calculations;

- A statement as to which M&V option will be used. If more than one option is used, the statement should include procedures to be sure savings from one option are not counted twice;

- A description of the exact data analysis procedures, algorithms and assumptions that will be used;

- A specification of the data that will be made available if a third party is to verify the reported savings;

- A notation of the budget and resources required for both the setup and ongoing M&V costs; and

- A format and/or specification of how results will be reported and documented.

Even a cursory review of this list makes it obvious that the roots of a quality M&V Plan rest upon a strong IGA. Ergo, a weak audit weakens the M&V Plan. Therefore, an IGA auditor must know how the audit information will be used in the M&V protocol. It follows that the auditor needs to understand why M&V is key to an effective program, the existing guidelines/options available for measuring and verifying energy savings, and the accepted M&V procedures.

THE CASE FOR M&V

In energy managers' offices across the country are decals, posters and mouse pads that read, "You can't manage what you can't measure." Interestingly, many come from the former Mobil Energy Management group. Even "big energy" knows measurement is key. And it's true: *You can't manage what you can't measure.*

Climbing utility bills and growing concerns about air pollution have underscored the economic and environmental value of energy efficiency. Unfortunately, all too often we do not know how effective our efforts are, or even if we were using our limited resources to the best advantage. M&V provides many answers, including:

- if energy was really saved;

- whether savings predictions of the IGA auditor were accurate;

- who should receive the credit (and payment) for any savings;

- if the most cost-effective measures were implemented;

- if the predicted environmental benefits were achieved;

- if a participating utility could, or should, receive rate adjustments, reimbursements, or other claimed benefits;

- the means for business managers and organization administrators to convince their boards or publics of their
  — environmentally responsible actions, or
  — sound business practices;

- if the specific measures the contractor implemented saved the energy, or if something else was cutting energy consumption in the facility or process; or

- if the savings being claimed by the contractor really came about from actions by the owner, which are not directly related to the project or the engineer's predictions.

It is, however, very easy to get caught up in the wonderfulness of M&V and move into overkill. An auditor can offer an important voice in the needed level of M&V and the most cost-effective procedures for getting the necessary information. An end user should never accept an M&V Plan at face value. The logic and benefit of each proffered task should be understood and questioned wherever warranted. The auditor can provide the end user valuable counsel in this regard.

Implementing an M&V protocol costs money. The more accuracy demanded; the greater the cost. However, M&V should never be allowed to become the major cost factor in a project. As a general "rule of thumb" M&V should rarely exceed (and usually falls below) 10% of the projected savings.

There are too many M&V programs that eat up capital that could have been better spent on the energy efficiency measures (EEMs) themselves. Unbelievably, some people have avoided doing energy efficiency work because a volatile baseyear made it very difficult to measure the savings accurately. As in any area of energy management, a little common sense goes a long way. If all parties agree the measure is beneficial and the savings are not guaranteed, then it is conceivable that an EEM can be installed without using any level of M&V.

M&V RULES OF THE ROAD

There are six basic M&V rules for end users and auditors. Auditors, engineers and energy efficiency consultants should be prepared to offer guidance as these rules are applied. The language in the following rules are directed at those offering technical support, especially the auditor, but the owner should be cognizant of these matters as well.

1.   Limit M&V to a cost level justified by the measure or project. M&V is always a matter of *cost vs. accuracy* and all parties involved in any project should determine collectively how much accuracy they are willing to buy. No more than 10

percent of the projected savings is a good guideline for M&V. Seldom does a project justify a higher M&V allocation.

2. If the purpose of a baseyear/baseline is as a reference point to which savings can be compared, then measures should be determined *before* the baseyear work is done. Baseyear information on the HVAC system is simply not needed if only lighting measures are going to be installed. Zealous M&V specialists too often encourage baseyear work for an entire facility at the outset. Later the auditor and the owner determine that only a few simple measures are warranted and the comprehensive baseyear work proves to be excessive and costly.

    If an owner wants comprehensive baseyear data for the energy management program or for other decisions, having the baseyear data for the entire facility may be very helpful. In such cases, this information will be part of the IGA. Acquiring the comprehensive databases, of course, needs to be weighed carefully against the value of such information. In such cases, however, the costs should not be entirely assigned to the M&V effort.

    Auditors should also be aware that some M&V specialists urge that they be involved at every step of energy efficiency project planning. Unless the M&V person is already on staff, this is a luxury that few organizations or projects can justify. Until measures are known, no detailed planning that would require an outside M&V professional's participation can realistically be undertaken.

3. If a performance contracting is involved (where money changes hands based on the savings determinations), allowing one of the parties of an agreement to do the M&V work is a judgment call. A potential conflict of interest exists when money changes hands based on the determined savings. Securing the services of a third party M&V specialist, however, is not always warranted. Some projects are just too small or too simple, such as lighting only, to warrant the trouble. For

large, complex projects third party validation may be a very good investment.

4.   The auditor and the owner should have a basic understanding of what M&V options are available and the respective strengths and weaknesses of each option relative to the measures under consideration. M&V options can, and should, be established as soon as approved measures are known. Criteria for selecting the options to be used should include;

   a.   project size,
   b.   measures selected,
   c.   consistency of usage, patterns,
   d.   type of savings documentation needed,
   e.   if payments are related to savings achieved,
   f.   instrumentation available and whether permanent installation is warranted/possible, and/or
   g.   accuracy required; cost of that accuracy.

5.   If an M&V specialist seems warranted, the owner should check the credentials and may ask for the IGA auditor's counsel in this regard. The International Performance Measurement and Verification Protocol (IPMVP)[4-2] has established a Certified M&V Professional program in cooperation with the Association of Energy Engineers.[4-3] This certification procedure helps bring quality and consistency to M&V work around the world and helps those seeking M&V counsel identify quality M&V Professionals.

6.   Your client, the owner, cannot abrogate his/her responsibilities to conduct oversight of the M&V program and may call on you to help. In that capacity as an auditor, you can offer guidance in establishing the customer's responsibilities, and helping them recognize that the M&V process starts with the identification of the measures and continues throughout the project. "Too little, too late" causes owners a lot of grief. 20/20 hindsight is painful and costly.

M&V GUIDANCE

Today, several accepted M&V approaches are available. The United States Environmental Protection Agency has done some work in this area. ASHRAE has developed its 14 guidelines, *Measurement of Energy and Demand Savings*, which are focused at a very technical level. By far the most widely recognized protocol is the *International Performance Measurement and Verification Protocol* (IPMVP) already referenced. The International Performance Measurement and Verification Protocol is regarded by most as the overarching document. Efforts have been made to have the IPMVP, or MVP as it is commonly called, compatible with the EPA and ASHRAE documents.

The initial steps for the MVP were taken in the United States by a cross section of representatives of the government, utilities, M&V companies, and the energy efficiency and performance contracting companies under the U.S. Department of Energy's sponsorship. With slight modification, this protocol can be used in any country.

As with any continuing work in progress, such as the MVP, regular revisions are expected. The most recent version was released in January of 2001 and can be obtained from the protocol's homepage *http://www.ipmvp.org*. The Protocol is now presented in two volumes and both are available at the web site. Volume 3, "Applications," dealing with M&V concerns related to new construction, renewables and water conservation, is expected to be released in late 2003 or early 2004.

The MVP has become the de facto protocol for M&V in performance contracting. Institutions, such as the International Finance Corporation, have found the Protocol beneficial and are incorporating it as a required part of new energy efficiency projects. The 1997 version has been broadly accepted in at least 25 countries around the world and translated into Bulgarian, Chinese, Czech, Japanese, Korean, Polish, Portuguese, Romanian, Russian, Spanish and Ukrainian.

The 2001 IPMVP version was developed with the help of

hundreds of organizations and is truly an international protocol with valuable input from experts from over 25 countries. The IPMVP Executive Committee of 15 members is charged with setting policy and managing the protocol development. It's international in make-up with members from Hong Kong, India, Brazil, Italy, France and Singapore.

MVP as a Work in Progress

Auditors need to be cognizant that the MVP is always a work in progress. Every new version of the MVP incorporates changes and improvements reflecting new research, improved methodologies and data acquisition procedures; so it is always in transition. To the extent that you, as an auditor or consulting engineer, are involved in the M&V procedures, you need to make use of the web site to view new and/or modified content, interim revisions to the existing protocol.

The IPMVP leadership welcomes comments from users. M&V specialists are encouraged to use the web site to review drafts as they are prepared. Continued development and adoption of MVP will involve increasingly broad international participation and management of the document as well as its translation and adoption.

M&V Options

The four options provided in the MVP are described very briefly below. For those who wish to have a little more depth of understanding regarding the options and their applications, additional details are available in the appendices. The material in the appendix is designed to aid those who wish to be more conversant regarding the options, but it is not adequate to perform the M&V tasks. Those performing M&V work need to use the IPMVP guidelines themselves.

Option A. Partially Measured Retrofit Isolation

Savings are determined by partial field measurement (some, but not all, parameters may be stipulated) of the energy use of the system(s) to which an energy efficiency measure (EEM) is applied, separate from the energy use of the rest of the facility. Measurements may be either short-term or continuous. This option involves the isolation of the energy use of the equipment/system affected by an EEM from the rest of the facility.

Option B. Retrofit Isolation

The savings determination techniques of Option B are identical to those of Option A except that no stipulations are allowed under B. Pre- and post- measurements are required. Savings are determined by field measurement of the energy use of the systems to which the EEM is applied, separate from the energy use of the rest of the facility. Short-term or continuous measurements are taken throughout the post-retrofit period.

Option C. Whole Building

Option C is often referred to as the "Whole Building" approach; however, this option can be used for a designated part of a building. Option C has also been called the "main meter" approach, and can be used to determine the collective savings of all EEMs applied to a specified part of the facility, which is monitored by a single measurement device. Short-term or continuous measurements are taken throughout the post-retrofit period. Option C usually relies on *continuous* measurement of whole-facility energy use and electric demand for a specific period before retrofit (baseyear) and *continuous* measurement of the whole-facility energy use and demand, post-installation. Measurements may be taken on a periodic basis if acceptable to all parties involved.

Option D. Calibrated Simulation

Savings are determined through computer-based simulation of the energy use of components of the whole facility. Simulation routines must be calibrated so they predict an energy use and demand pattern that reasonably matches actual energy consumption. This option requires considerable data input, is costly and requires a fairly large project to justify the cost.

Option Use Cautions and Limitations

None of the above options are apt to provide irrefutable data. Auditors should always stress to the owners that caution in using M&V data is wise. Under most M&V procedures, the level of accuracy at best is ±10 percent.

To the extent that the M&V procedures gauge the auditor's work, the IGA auditor should be aware that caution is warranted in the use of Option D, as this option typically requires considerable skill in calibrated simulation and considerable data input. Unless implemented correctly, the results will not be reliable.

The two isolations options, A & B, do not provide for the interaction of the measures. When one measure, such as lighting, affects another aspect of the energy systems, such as heating and cooling, these M&V options will not reflect the interaction. Relying solely on Options A or B can give a false impression of the total amount of energy saved. Retrofit isolation options offer valuable information regarding a specific measure, but they do not offer a net savings benefit for a multi-measure project.

Calculated savings by isolated measures are not additive. In the late 70s, engineers sometimes provided energy savings calculations by measure with an implied total benefit. This approach has often been described as "pumping oil back into the truck." Far too often the projected savings added up to more fuel than the owner purchased.

As every capable auditor knows, every time a measure is implemented, the potential savings "pie" shrinks proportionately. To disregard this factor leads to exaggerated savings—and, if payment changes hands based on the savings of each EEM, your client may end up paying for savings not actually realized. In making their calculations, quality engineers assume the previous measure(s) have been implemented in predicting the savings of succeeding measures. It cannot be overstated that any effort to compare M&V work with the engineer's calculations needs to follow the same procedure, and rank the measures in the same order.

Option C does not sort out the measures; so attributing the savings to a given party, such as an ESCO, for the work done is not always possible. Under the Whole Building approach, the owner's actions can wipe out the ESCO's gains. Or, conversely, the ESCO can take credit for the end user's energy saving actions.

Other than cost, the greatest limitation in using Option D is the quality of the input data and the qualifications of the person performing the work.

Changes in the IPMVP 2001 Version

Since the IPMVP has been around for a number of years, auditors may find it helpful to consider the changes in the newer version compared to its predecessor, the 1997 version.

As noted earlier, the 2001 version is now presented in three volumes, which should make its use and the option's selection much easier.

Volume I. Concepts and Options for Determining Savings

This volume is largely drawn from the 1997 edition. Options A and B have been significantly modified in response to reactions received to earlier editions. These changes now include required field measurement of at least some variable under Option A, and all aspects under Option B. Helpful examples of each M&V Option have been provided in Appendix B of the document. Former sections of the 1997 version; i.e., new buildings, residential and water efficiency have been moved to Volume III.

Volume II. Indoor Environmental Quality (IEQ) Issues

Guidance as to the relationship of energy efficiency to indoor environmental issues is provided. The focus is on measurement issues, project design, and implementations practices associated with the maintenance of acceptable indoor environmental conditions within the context of an energy efficiency project. The discussion also relates to elements of M&V and energy performance contracts.

Volume III. Applications

At one point, it was hoped that Volume III would be published in early 2001, but it was decided to make more extensive revisions to portions of the document. In particular, a new com-

mittee was constituted to improve the new construction guidelines. Their work has been approved and now appears on the
organization's website: ipmvp.org. It will eventually be included
in Volume III. Other areas to be addressed in Volume III will include water efficiency and renewable energy.

New features in the version published in January 2001, include:

•   Adherence language with specific steps to follow for a con-
        tractor to claim adherence to the MVP;

•   Guidance in using the basic approach and preparing a good
        M&V Plan;

•   Clarification of the term "stipulation" and its acceptable use
        within the MVP framework;

•   How to establish the baseyear and further guidance regard-
        ing "adjustment" to bring energy/demand baseline to the
        same set of conditions for pre- and post-retrofit;

•   A new Volume II on improving indoor environmental quality
        while implementing energy efficiency measures; and

•   Greater internal consistency and clearer directions for using
        the Options, particularly Option A.

IPMVP Adherence

Many contracts and M&V plans specify that once the measurements are known, the IPMVP protocol will be followed. To be
sure such claims are justified, special adherence language was
inserted into the MVP 2001 version. It can be found in section 3.5
of the document. Because procedural compliance can be so important, it is recommended that auditors be familiar with what constitutes compliance with the IPMVP.

MEASUREMENT DEVICES

Owners often expect their engineers to provide guidance in the appropriate use of various measuring devices. The various instrumentation devices described below have differing benefits and can be applied with success if the purpose and limitations are clearly understood.

Portable Data Loggers (PDL)

One of the simplest measurement instruments is a portable data logger. About the size of a pocket calculator, these battery-operated devices most often record hours of operation, although there are a few that record temperature and consumption.

The use of PDLs is normally limited to fixed load, variable hours of operation measures. For example, PDLs are often used to record operating hours of light fixtures. Retrieval of the data is often a manual process, so human error can be a factor. Costs associated with using a PDL are dominated by the labor required to place the devices and retrieve the data. As technology has improved, more complex PDLs have been appearing, which can record hours of operation according to a utility rate schedule with data retrievable via a laptop computer. These devices can be cost effective for short-term measurements to verify estimated hours of operation in a facility.

The biggest drawback to these devices is that most of them do not permit analysis of operation by billing rate period. Energy rates are complex, and it is often necessary to determine when savings occurred in order to accurately assess the value of those savings. The 'snap shots' generated by PDL's will not be sufficient for this purpose.

Permanent Microprocessor-based
Universal Data Recorders (UDR)

These devices are multi-channel microprocessors that accept industry standard signal inputs to measure almost any variable in real time and record values for later retrieval over an auto-dialed telephone link. The great benefit of these devices is that the values are recorded at user definable intervals and permit peak coincidence analysis.

UDRs are the preferred device for most contracts where payment is savings based. They are designed specifically for use in M&V applications and, importantly, have integrated security functions that ensure maintenance of the integrity of original data. An independent assessor can access the data and ensure that the savings reports are accurate.

The universal nature of the signal inputs, combined with the interval recording capability allows UDR devices to meet the needs of all M&V protocols. To date they have been used primarily in lighting applications to verify hours of operation; however, more complex applications of UDR devices are becoming more common, which include variable load, variable hours of operation and weather dependent loads, such as HVAC systems.

One important feature of a UDR is its ability to measure electrical load. Most of the devices that are commercially available today require a separate kilowatt-hour meter with a pulse output.

Building Automation Systems (BAS)

Another type of measurement device is the BAS. These systems may be used to log control variables to determine the energy savings that have resulted from the control algorithms. There are several problems with the use of BAS for M&V. The cost per point for a BAS system is often more expensive than that of a UDR. Typically, one has to pay for the development costs of the control functions in addition to the measurement features. Security may also be a problem. BAS are designed to record the performance of the control algorithms for tuning purposes and rarely include security features on the data. The logging capabilities of BAS are also very limited. In order to meet the needs of measurement protocols compromises are usually required, often sacrificing calculation of peak coincidence.

Software

Both of the major international circuit breaker companies, Cutler Hammer (a division of US-based Eaton) and Square D (a French company), have been active in developing software, which can be attached to the circuit breaker to measure consumption. Cutler Hammer has been particularly attentive to M&V applications. This software can read energy down to machine level and is valuable in dynamic situations where attributing savings is difficult.

Using such software is particularly helpful where performance contracting payments are based on specific pieces of equipment. It is usually less costly than sub-metering and can be used proactively to manage energy as well as passively to document previous savings. This software is also valuable in measuring other electrical qualities of interest, such as harmonics.

Software is increasingly available from other companies, such as Internet Energy Services, which are very adaptable to specific measurement needs and has resolved many of the open protocol issues. Sixth Dimension has also done work in this area, but has undergone major changes in its leadership and is currently struggling to fulfill its commitments.

COMMISSIONING AND THE M&V FIT

The purpose of commissioning is to ascertain that the design criteria have been met and to determine that all installed equipment is operating correctly according to specifications. Commissioning should be part of the overall plan (See Chapter 10). Since commissioning is a performance verification procedure, it can also serve as a key first step in the measurement and savings verification process.

If an owner is paying for commissioning, they should not be paying for M&V procedures that do exactly the same thing. If part of your responsibility to the owner is to offer broader guidance, be sure that the M&V plan accepts that commissioning has been done and overlapping tasks are eliminated.

WHEN M&V JUST ISN'T WORKING

The best laid plans of mice and men... happen in M&V, too. When it does, the first step is for the parties involved to sit down calmly and discuss it. If a good communications plan has been effectively imple-mented, this is the time when it really pays off. Too often, some rather hysterical circular finger point-ing gets started and nothing is resolved.

First of all, own-ers need to be realistic about the accuracy of the M&V procedures offer. Some options only offer ±10-20% ac-curacy. No one should expect more than the plan promised to de-liver.

Experience sug-gests that if the M&V is not yielding the ex-pected results, the most likely culprit is the original baseyear calculations and/or the baseline adjust-ment provisions. Next on the list is a search to determine if any

WHO'S TO BLAME

modifications in the facility or procedures, which affect energy consumption, have been made and not reported. An examination of operations and maintenance practices and any changes that may have been implemented could be crucial.

After the more sweeping possibilities have been eliminated, an examination of the individual EEMs recommended in the IGA and the associated calculations should be made. If whole building Options C or D have been used, then isolating each suspected EEM and measuring consumption might be warranted.

The M&V procedures are designed to tell you how much the EEMs saved, or didn't save. It is not within the M&V report's purview to tell you why the measure did not perform as predicted. Some "M&V specialists," who banner their talents by telling horror stories about performance contracting or M&V, often mix up the two procedures. M&V is one thing; an engineering analysis of why it didn't work as predicted is something else. It is possible for the same person, or firm, to perform both services; but it is in the best interests of the owner to keep the processes separated. Someone selling engineering analysis often has a conflict of interest in reporting the M&V results.

If all else fails, it is just possible that the M&V program itself has not done the job. The 8 steps "common to all good savings determination" as outlined in IPMVP's Section 3.2 may not have been followed. The M&V specialist may not have performed adequately. Getting a *certified* M&V Professional is an important precaution.

If uncertainty exists regarding the M&V specialist's work, the least expensive solution may be to engage another M&V professional to run a check on the program. Selecting a qualified M&V Professional in the beginning is, of course, more satisfying and often cheaper. In addition to the certification credentials, a few questions may help your clients be sure they are getting the best professional for *their* operations; so urge your client to ask for:

•   observations regarding the facility/process, contemplated measures and M&V needs;

- their thoughts on when M&V Professionals should get involved in the project;

- typical costs of M&V in relation to savings projections or the construction costs;

- recommended M&V option for specific measures;

- the range of errors that can be expected with their recommended approach;

- appropriate measuring devices for the EEMs being considered and how the determination as to their use is made;

- their recommendation on how M&V results should be reported; and

- what they perceive as the greatest weaknesses in M&V work, particularly in relation to the EEMs being considered in the project.

Answers should be weighed carefully against criteria set forth in this chapter.

In summary, M&V protocol may seem complex, but it has a logical order. With a little homework, the process can easily be mastered. For IGA auditors, a clear understanding is a must and certainly worth the effort.

In the final analysis, anything can be measured and any savings can be verified if one spends enough money. There is an inclination by M&V specialists to overplay M&V aspects, which in turn places a burden on the project. If M&V becomes too costly, the measure will no longer make economic sense. It is always a question of cost vs. accuracy. The owner, contractor (and perhaps the financier) should sit down and agree on what constitutes a *reasonable* level of accuracy. The bottom line is: Just how much accuracy can the project afford and the owner/contractor/financier justify?

M&V procedures and a quality IGA are inextricably linked. An M&V program relies on strong baseyear data and baseline adjustment procedures. These data are rooted in the audit. The quality of the data is dependent on a solid audit. At the same time, our best gauge of an auditor's abilities is predictive consistency. Further, predictive consistency can only be assessed with measurement and verification procedures that adhere to clear, reliable guidelines.

One final harsh admonition: End users, who wait until there is a problem to get involved in the M&V effort, probably deserve what they get. Unfortunately, the "buck" doesn't stop there. It is all too easy to pass the blame along—and auditors make a convenient target. A little advanced self-preservation (hopefully in writing) is a good precaution for the auditor and provides good counsel for the client.

References

4-1   The term, baseyear, used in this text refers to the historical record, which distinguishes these data from the annual adjusted baseline. This usage is consistent with the leading M&V protocol document, the 2001 International Performance Measurement and Verification Protocol referenced later in the chapter.

4-2   Information on the International Performance Measurement and Verification Protocol can be obtained by visiting the organization's web site: ipmvp.org.

4-3   More information on the Certified M&V Professional's program can be obtained by contacting The Association of Energy Engineers, Atlanta, Georgia; phone 770-447-5083.

Chapter 5

The IAQ Fit

TO FULFILL the indoor air quality (IAQ) expectations in an investment grade audit (IGA), an auditor must recognize significant air quality issues and understand the relationship between IAQ and energy efficiency efforts. Whether the systems you were hired to analyze are the source of the air quality problem or not, they are frequently involved in the distribution of those problems. The reality is that if those systems are involved, then so are you.

IAQ AND ENERGY EFFICIENCY

The real relationship between energy efficiency (EE) and indoor air quality (IAQ) should be known to every IGA auditor, for this is an emotionally laden area rampant with misinformation. Energy efficiency has often been the focus of IAQ fears and your recommendations may come under unfounded attack. The fact that IAQ fears related to energy efficiency are usually unwarranted does not prevent them from spreading throughout an organization. An auditor, therefore, must be well-armed and well-informed to respond to the near hysteria that sometimes can occur when occupants are advised energy efficiency measures (EEMs) are going to be instituted. This chapter explores the true relationship of IAQ and energy efficiency.[5-1]

We are still haunted today by articles starting in the early 1990s about "sick building syndrome" where the second or third paragraph invariably mentioned the energy crisis of the 1970s, the resulting tight buildings, and the beginnings of our IAQ woes. Such articles regularly left the impression that, as energy prices soared in the 70s, nasty owners and facility managers tightened buildings to save dollars and left occupants sealed in these boxes gagging on stale pollutant-laden air.

To deal with this heritage, we need to first ask where did we get the idea that the energy efficient building was at fault? For an answer, we need to reach back to the original test of "fresh" air needs: the canary in the mine shaft. If the canary died, we knew we needed more fresh air.

Or, we can go back to a time when the old, leaky, drafty building was supreme. Or, using energy efficiency culprit thinking, if we can't go back to those old leaky buildings, let's open the windows and turn up the fans. Surely, we need more *fresh* air in those buildings. For over a decade, ventilation disciples have almost convinced us that's the way to go. Open the window! Air will just naturally get better.

But has it? Will it? The answer, unfortunately, is not necessarily.

The Fresh Air Answer

Natural air sounds so wholesome, but outside air can be heavily polluted. It's easy to think of a number of places; e.g., airports, bus stations, pulp and paper factories, etc., where the air outside can be much worse than the filtered air inside. In such cases, natural ventilation could be a disaster.

Ask any hay fever sufferer about opening the windows and letting all that wonderful fresh air in. And all the pollen with it.

Or, we can look at it from another perspective: What happens inside a building when a window is opened? What seemed like a good idea can cause a stack effect. It can draw in pollution from traffic on the streets below. Or, opening that window could create negative pressure in the basement. Now, if the building has radon problems, the increased cross ventilation could cause even more radon to be drawn into the work area.

Ventilation is not always the answer.

To properly define the relationship between IAQ and energy efficiency, we need to make that statement even stronger. Ventilation is seldom the best answer. Certainly, it's an expensive one.

Is EE Really the Villain?

Popular assumptions hold that energy efficient "tight" buildings don't provide enough ventilation. Ergo, the problems can be rectified by more ventilation. In the following discussion, for "ventilation" read *energy*.

For years, the American Society of Heating, Refrigerating and Air-conditioning Engineers (ASHRAE) titled its Standard 62, "<u>Ventilation</u> for Acceptable Indoor Air Quality." [underscore supplied]. To the uninitiated, that sure sounds like ventilation will deliver acceptable air. At the very least, the title sounds like ASHRAE has given its blessing to ventilation as THE mitigating strategy.

Make no mistake, we firmly believe that ASHRAE's 62 Standard has brought relief to many people who would otherwise have suffered from indoor pollutants. At the time the standard was developed, it was very difficult to determine what some of the pollutants were. Back then, increased ventilation probably gave relief to occupants in an era when no one was quite sure what else to do. On the downside, however, ASHRAE 62 has done a disservice to many people, because it has given credence to the idea that ventilation was the answer to IAQ problems.

The other document feeding the "energy culprit" thinking was a pamphlet put out by the National Institute for Occupational Safety and Health (NIOSH) at a time when so many were trying to come to grips with IAQ. The document said that 52 percent of IAQ problems NIOSH had analyzed were due to "inadequate ventilation," which unfortunately was translated into *inadequate outside air*. A closer look at those *inadequate ventilation* conditions revealed that it included such problems as ventilation effectiveness (inadequate distribution), poor HVAC maintenance, temperature and humidity complaints, filtration concerns, and inadequate outside air. With what we know now about the contributions of poor HVAC maintenance to IAQ, and the recognized level of temperature and humidity complaints, the only surprise is that the figure was not higher than 52 percent.

NIOSH later observed that the 52 percent figure was based on soft data. To the extent, however, that it represented primary problems in the investigated buildings, the NIOSH findings also imparted another critical piece of information that over the years has been typically overlooked: *48 percent of those problems could NOT be solved by ventilation.*

The Real IAQ/Energy Efficiency Relationship

A relationship between IAQ and energy efficiency does exist and it is borne out by research. First, we can trace our problems back to the 70s, but not to tight buildings. Numerous surveys have shown that when utility bills started climbing in the 70s, the first place owners and facility managers looked to find money to pay those bills was in the maintenance budget. This was especially true of institutions on rigid budgets, such as public schools and hospitals.

As utility bills continued to climb through the years, those institutions progressively cut deeper into maintenance. The result:

their deferred maintenance bills are staggering. The public schools' deferred maintenance bill surpassed $100 billion some time ago.

At the same time maintenance was being neglected, owners bought more sophisticated energy efficient equipment. Unfortunately, operations and maintenance (O&M) training wasn't sufficient. Sometimes the training wasn't offered when the equipment was installed. More often, it was offered, but then a turnover in the O&M personnel lost that training advantage as the new staff didn't receive the necessary training. With this in mind, consider that in the early 90s the major IAQ investigative groups found that up to 75 percent of IAQ problems could be traced to inadequate maintenance. When we pair this and the $100+ billion in schools' deferred maintenance, it is not too surprising that one study found that 70 percent of all schools buildings have IAQ problems.[5-2]

It is important that we don't minimize the fact that poor quality indoor air can have a major impact on the work environment. As a case in point, let's take a closer look at educational

---

Schools and IAQ

Children breathe a greater volume of air relative to their body weight; so they may be more sensitive to indoor contaminants. Occupant density in a school is also a factor. Schools' population density is typically four times that of an office building. We should underscore these factors with a realization that the law requires children to be housed in these facilities (70% with IAQ problems) for at least five hours a day. Further, there is mounting evidence that the quality of a school's physical environment affects educational achievement, so concern is heightened further.

A 1995 study by the General Accounting Office reported that ventilation, indoor air quality, temperature (heating and cooling) and lighting are among the leading unsatisfactory environmental conditions in school buildings.

The EPA has devoted considerable time and expense to researching school IAQ matters. More information can be obtained from the agency.

facilities and IAQ.

Most auditors would agree that one of the major goals of an audit is to provide the client with a healthy, safe and productive environment as cost-effectively as possible. If, indeed, that is your goal, then IAQ and energy efficiency are not adversarial concepts, but must go hand in hand.

Consider what will happen as energy prices go up—and ultimately they will. If we try to put IAQ and energy efficiency at loggerheads, IAQ will lose. Environmental concerns, higher energy prices, and the unnecessary waste of our limited energy resources make increased ventilation, at the very least, a costly answer. Auditors need to work with contractors and clients for a quality indoor environment as well as an energy efficient one.

Ventilation should not be the preferred treatment for IAQ problems. A closer look suggests that it never has been! The Environmental Protection Agency (EPA) has been telling building owners, facility managers and engineers for years that the best mitigating strategy is *control at the source*.

In the 70s and 80s, outside air intake was reduced to alleviate the high energy costs of conditioning and moving air around. With less outside air, owners and occupants suddenly became more aware of the contaminants that had been there all along. In fact, in the interim, printers, copiers, etc. had introduced even more contaminants. Less outside air meant greater concentrations of these contaminants.

Since reduced ventilation was a fairly standard remedy in the 70s, it isn't surprising that the knee-jerk reaction to possible IAQ problems has been to increase ventilation.

Losing Ground

Ironically, increased ventilation may cause us to lose ground on indoor air quality. Consider what usually happens to humidifiers/dehumidifiers in the specs when faced with a construction cost overrun. Humidifiers and dehumidifiers have historically been at the top of the list of measures that can be cut when construction costs exceed the expected.

Today, without those humidifiers or dehumidifiers, it's very hard to correct the negative impact increased ventilation has on relative humidity. Ask the folks in Florida what happened when they increased ventilation to meet the requirements of ASHRAE 62-89. They'd be happy to show you the mold.

And in our northern tier states we've had problems with air that is too dry. Heating up additional outside air just exacerbates the problem. With more than 50 years of data on respiratory irritation and illness due to dry air, creating drier air has not been the answer. Too often increasing outside air without considering the impact on humidity can change a "maybe" IAQ problem into a definite one.

The "dilution delusion" has also mislead us into thinking ventilation has cured the problem. Visualize for a moment all those airborne contaminants as a bright purple liquid flowing out of a pipe in an occupied area. What would happen if you recommended to your client that they just hose the area down each morning? Just because we can't see air pollutants does not mean they're not there—even if they could be diluted to a light shade of purple.

There is still a lot we don't know about chronic low level exposure to some contaminants. Further, we are facing a very real possibility that we'll look back in a couple decades and see solution by dilution as nothing but delusion. A very serious delusion. Consider your legal position today if you had been an ardent advocate of asbestos 35 years ago.

The Dangers of "Assume"

We've all heard about the dangers of making assumptions. Yet, people repeatedly make some gargantuan assumption about outdoor air as a good IAQ mitigating strategy.

There is a rather blatant assumption that increased outside air will reach building occupants. Yet, studies of ventilation effectiveness have repeatedly found ventilation designs that short circuit the air flow. When you audit a building, look for the diffusers. Where are they? How does the air leave the room? Where does it enter the room? Increasing outside air may cause a nice breeze across the ceiling, but it may do little for the occupants.

The "natural" outdoor air fetish may have also caused us to bring in outside air when recirculated cleaned air could have been better. Filtration and air cleaning may be a much better answer for some of your clients than increased ventilation. And we can't ignore the fact that all that fresh air has to be conditioned and moved through the facility. Somehow we've lost sight of the ramifications of burning fossil fuels unnecessarily to condition all that additional air we bring in. Anyone hear of the Kyoto Protoctol?

Probably the greatest mistake in all this is what we call the Pogo Fallacy: "I have met the enemy and he is us." Us filthy, dirty, polluting occupants who are known for shedding skin, hair, and other disgusting things. The typical ventilation requirements have historically been based on cubic feet per minute per occupant. This kind of people-pollution thinking has its roots in the Dark Ages when people took a bath once a year whether they needed it or not. If you, or those questioning your EE recommendations, think body odor and cigarette smoke are the only causes of IAQ problems, then it's time to re-think that premise.

Today, human pollution is but a small part of the problem. Yes, we are still those breathing, shedding creatures, but we've been joined by other contaminant-making things. When we try to satisfy our IAQ needs through cfm of air per occupant, we forget about all the other pollutants we've introduced through the years.

Ventilation per occupant just cannot do the job if pollutant sources other than people dominate an area. Or, if we have low occupancy situations. Consider carpet that has been flooded. Unless disrupted, the bioaerosols being emitted from the carpet are the same if there are 3 or 300 in the room. Or, an office setting where the printer and copier are grinding out work at the same pace whether there are 2 or 20 in the room. For years we lived with the infamous Table 2 of ASHRAE 62-89; however, even with its exceptions, it never addressed occupancy situations below those assumed in the table.

Then, there is the carbon dioxide (CO_2) issue. Because CO_2 is a measurable pollutant, many designers and building owners are installing monitoring systems throughout their facilities. Obviously, being able to quantify the level (described in parts-per-million, or ppm) is a good thing. Especially since ASHRAE, using CO_2 as a surrogate for contaminates that cannot be readily measured, has given us a relative goal, or maximum allowable ppm, that allows us to measure the relative contaminate level.

The problem is that the 1000 ppm indoor allowable level presented by ASHRAE 62 and modified slightly by an addendum really strives to achieve the real goal of not having indoor readings exceed outdoor air levels by more than 700 ppm. In other words, the original 1000 ppm indoor level presumed an average 300 ppm outdoor air CO_2 level. So those facilities without outdoor CO_2 monitors are only telling half the story, and areas with less than 300 ppm outdoor readings could be providing a false confidence to building owners whose indoor level is more than 700 ppm above the outdoor level.

IAQ FUNDAMENTALS

Understanding various aspects of IAQ is a must for an IGA auditor. Ventilation codes, for example, are absolutely critical. Assessing and reporting any expected changes in IAQ conditions as a result of recommended energy efficiency measures is essential. Other aspects are more subtle, but offering a higher quality audit makes them important considerations. And if the IGA leads to a Master Plan, as we recommend, then full consideration of a facility's IAQ needs becomes key.

Many facets of IAQ management are not as complicated as they may seem. IAQ investigative procedures over the years have shown that roughly 80 percent of IAQ problems can be traced to inappropriate exhaust or inadequate maintenance. Most of these problems can be identified in a walk through audit. It is incumbent on an IGA auditor to be able to identify such problems and recommend steps to rectify the situation.

At no time, should an auditor guarantee the resulting quality of the indoor air. You can say your work is designed to reduce pollution emissions. You can propose to find common sources of pollutants, etc. But in any dynamic situation other factors over which you have no control can nullify your efforts. Consider that at the dawn of the new millennium we were still finding about 400 *new* volatile organic compounds (VOCs) per year.

Standards and measurement procedures are not in place to document all aspects of IAQ. There is not an objective level of IAQ you could promise even if you wanted to. *And you don't want to!*

An IGA auditor should be very familiar with ASHRAE 62 provisions and local codes. Further, the conditions resulting from any recommended energy efficiency (EE) changes should comply with these codes and standards. An owner has the right to expect such compliance and to be confident that the audit has identified the more common IAQ problems that might exist in a facility.

Compliance with existing codes and standards has been known to increase energy consumption. When this is the case, the IGA report should very clearly state this impact on efforts to reduce energy consumption.

It is absolutely critical, however, that you remind your clients, up-front, in situations where IAQ problems exist that 20 percent of the problems will not be identified in a walk through audit. Of that 20 percent, roughly 10 percent can be found through extensive (and expensive) testing and analysis by IAQ professionals. Unfortunately, despite our current technical and scientific capabilities, nearly 10 percent of our IAQ problems still remain undetected.

The specialized knowledge needed to identify many HVAC related IAQ problems as well as other potential IAQ concerns go beyond the purview of this book. Auditors, who wish to pursue this area, are encouraged to read *Managing Indoor Air Quality*[5-3], particularly Chapters 4, "Investigating Indoor Air Quality Problems: How to Find Out What Went Wrong;" Chapter 7, "HVAC: The Heart of IAQ;" and Chapter 9, "Operations & Maintenance:

An Ounce of Prevention." If you are going to provide management consultation to you client, Chapter 10, "Management Procedures: The Soft Side of IAQ Success," can provide you some valuable guidance.

SAVING ENERGY AND LOSING PRODUCTIVITY?

Some "tight building" alarmists have exclaimed over increased absenteeism and lost productivity. In effect, many have been saying, "What a stupid idea—saving energy—for a few pennies... and losing big dollars to do it. That crazy auditor's recommendations are ridiculous!"

It's past time to clear the air. To prove this correlation, one would have to show that a majority of energy efficient buildings have poorer air quality and lower productivity. Furthermore, you'd have to be able to establish an inverse relationship, showing that the more energy efficient a building becomes, the greater the absenteeism and lost productivity. The work of Dr. Joe Romm[5-4] has shown just the opposite to be true: the more efficient a facility; the more productive it is.

The truth is that there's nothing wrong with a tight building—provided it is well-designed and well-maintained. Of course, tight buildings will more readily reveal professional errors. Certainly, tight buildings are less forgiving of poor maintenance. But a well-designed, well-maintained tight building can provide energy efficiency AND quality indoor air.

Simply put, an investment grade audit cannot ignore air quality implications when making energy efficiency recommendations. Aside from the obvious humane issues, an auditor has responsibilities to the owner and some serious legal concerns. An auditor, who fails to include IAQ considerations in the audit, does so at his/her own peril.

Occupants worry more about their work environment than they do about energy. Owners worry more about productivity than they do about energy. It is the IGA auditor's job to find ways to make energy efficiency enhance building environment... we call that the "IAQ fit."

References

5-1   The material presented in the remainder of this chapter is adapted from an article, "Clearing the Air: Indoor air Quality and Energy Efficiency," by Shirley J. Hansen, which originally appeared in *Energy and Environmental Management* magazine and was reproduced in *Contracting Business*, October 1995. The article was based on a keynote address for an ASHRAE Indoor Air Quality Conference.

5-2   Dorgan Associates. *Productivity Benefits Due to Improved Air Quality*. 1995. p. 37.

5-3   Hansen, Shirley J. and H.E. Burroughs, *Managing Indoor Air Quality, Second Edition*. The Fairmont Press, Lilburn, Georgia. 1999.

5-4   Romm, Joseph J. "Worker Productivity Rises with Energy Efficiency," *Strategic Planning for Energy and the Environment*, Vol. 14, No. 3, 1995. Also see Dr. Romm's book, *Lean and Clean Management*, available from Kodansha International, 114 5th Ave., New York, NY 10011.

Chapter 6

Financing Issues
And the IGA

THROUGHOUT THIS BOOK, we have repeatedly made the case that *energy efficiency is an investment, not an expense*. It follows, therefore, that there is a close relationship between the quality of the audit, which should serve as an investment guide, and financing considerations. It is not an exaggeration to say that the IGA and the project's financing are, or should be, inextricably linked. The client, and the means for financing the project, dictate the investment grade audit's parameters and procedures.

Money issues surround energy efficiency projects and impact the IGA process. The measures to be recommended, for example, are typically constrained by the established aggregate payback. And payback, of course, is a function of price. The costs of equipment acquisition and installation are also determining factors. The level of risk assigned to the contractor carries with it the associated costs of accepting, managing and/or mitigating such risks. If the audit is being done for an energy service company (ESCO), rather than an owner, issues related to the guarantees must be factored in. And the list goes on.

The mention of risk within a financial context deserves special attention; for, as mentioned in the previous chapters, money always follows risk. Very simply, owners regard risk as an expense item. Conversely, firms, who meet customer needs through the implementation of an IGA can turn risk management into a profit center—a very lucrative profit center—if they can effectively manage those risks. That's a big "IF," which is treated more fully in the next chapter, "Working Risk into the Mix." Suffice it to say here that the IGA is a critical risk management tool; and, therefore, has a major impact on financing considerations.

As a firm becomes more adept at managing and mitigating risks, it will find that risk management/mitigation can become a core business, but only if the firm does not become a risk "pass-though" channel to the insurance industry. The insurance industry, after all, has risk management as its core business. Instead, those who wish to capitalize on risk as a core business need to explore self-insured procedures, or the use of vehicles, such as performance assurance policies.

Performance assurance is far more cost-effective than insurance as the assurer and the client enter into a partnership. The mechanisms available allow clients to carry the degree of self-insured levels they are comfortable with. Further, the client can share in the proceeds from that portion of the financial protection not exercised as set forth in the agreement.[6-1]

To further explore the relationship of money to an IGA, we have separated the money facets related to project-based money matters into one category and those related to project financing *per se* into another.

BASIC MONEY CONCEPTS

To correctly calculate the value of a project, there are some basic money concepts that must be mastered, including life cycle costing, time value of money, discount rates, etc.

Life-Cycle Costing
Most auditors are aware of life-cycle costing procedures and the concept's relevance to a more accurate calculation of return on investment. Life-cycle costing (LCC) presents the net benefit of all major costs and savings for the life of the equipment discounted to present value. A building design or system that lowers the LCC without loss in performance can generally be held to be more cost-

effective. Other considerations, such as the calculation of present worth, discounting factors and rates, and LCC in new design, need detailed analysis and are more fully discussed in manuals and texts devoted to the topic.

Simple payback, at best, is a crude "first cut" determination and investment grade auditors should be aware of its limitations as well as the value of LCC. The simplest LCC formula is:

$$LCC = I - S + M + R + E$$

Where,

$$
\begin{aligned}
LCC &= \text{Life-cycle costing} \\
I &= \text{Investment costs} \\
S &= \text{Salvage value} \\
M &= \text{Maintenance costs} \\
R &= \text{Replacement costs} \\
E &= \text{Energy costs}
\end{aligned}
$$

LCC is obviously a more time consuming and costly procedure than simple payback calculations, but for larger measures and projects it is certainly warranted.

Time Value of Money

Money changes value with time. A simple question will put this idea into perspective: "Would you rather have us give you $1,000 today, or a year from now?" The astute will answer that they would rather have the money today. And they are not basing it only on the ol' "bird-in-the-hand" philosophy. They will tell you that they can invest that $1,000 and earn interest on it. If the interest rate were 10 percent, that $1,000 would be worth $1,100 a year from now. Conversely, if they wait a year to get our money, they have, in effect, lost the gains they might have achieved from interest and consequently that future $1,000 is worth closer to $900 at the present time.

To turn the time value of money into audit considerations, consider the following three options:

$1 million investment with a 1-year payback
$2 million investment with a 2-year payback
$8 million investment with a 8-year payback.

Keeping in mind that each option will yield $1 million per year, what would be the best investment? Without discounting future payments, the options would appear to be equal with regard to cash flow. Once discounting enters the picture, the answer, of course, is that since future payments decline in value, the first option is best.

As long as time is involved, there will be a discount rate. To get an idea of how this will impact an audit, look at the discount table on the following page[6-2] and decide how much the owner will get back *in today's dollars* for the second option above, and then for the third option. It will become readily apparent that the longer the payback period; the greater the discount rate impact. The higher the interest rate; the greater the decline in dollar value.

The discount rate is used to establish a net present value (NPV) for a project and should be incorporated in all IGA payback calculations. The difficulty comes in trying to project future interest rates. This is a risk that is hard to quantify. The best sources of assistance in making these determinations are the customer's financial officer or the project's financier.

All of this, of course, presupposes that some level of inflation exists. The magnitude of the problem will change if there is stagnation or deflation; however, as long as there is interest, there will be a decline in the value of money over time.

Project Financing

The cost of money, even its availability, depends upon the risks associated with a given project. A strong IGA that clearly has addressed potential risks is reassuring to financiers. A major function of an IGA, therefore, is to present the identified risks, document the possible management/mitigating strategies and assesses the total impact on the project.

As discussed at the end of this chapter, creating a "bankable" project depends heavily on the information provided in an audit.

Time Value of Money...

Formula: Discount Factor = $1/(1+i)^n$

Where: i = Discount rate in decimal form
         n = period in number of years

DISCOUNT RATE PERCENT

PERIOD (YEARS)	0%	2%	4%	5%	6%	8%	10%	12%	14%	16%	18%	20%	22%	25%
1	1.0000	.9804	.9615	.9524	.9434	.9259	.9091	.8929	.8772	.8621	.8475	.8333	.8197	.8000
2	1.0000	.9612	.9246	.9070	.8900	.8573	.8264	.7972	.7695	.7432	.7182	.6944	.6719	.6400
3	1.0000	.9423	.8890	.8638	.8396	.7938	.7513	.7118	.6750	.6407	.6086	.5787	.5507	.5120
4	1.0000	.9238	.8548	.8227	.7921	.7350	.6830	.6355	.5921	.5523	.5158	.4823	.4514	.4096
5	1.0000	.9057	.8219	.7835	.7473	.6806	.6209	.5674	.5194	.4761	.4371	.4019	.3700	.3277
6	1.0000	.8880	.7903	.7462	.7050	.6302	.5645	.5066	.4556	.4104	.3704	.3349	.3033	.2621
7	1.0000	.8706	.7599	.7107	.6651	.5835	.5132	.4523	.3996	.3538	.3139	.2791	.2486	.2097
8	1.0000	.8535	.7307	.6768	.6274	.5403	.4665	.4039	.3506	.3050	.2660	.2326	.2038	.1678
9	1.0000	.8368	.7026	.6446	.5919	.5002	.4241	.3606	.3075	.2630	.2255	.1938	.1670	.1342
10	1.0000	.8203	.6756	.6139	.5584	.4632	.3855	.3220	.2697	.2267	.1911	.1615	.1369	.1074
11	1.0000	.8043	.6496	.5847	.5268	.4289	.3505	.2875	.2366	.1954	.1619	.1346	.1122	.0859
12	1.0000	.7885	.6246	.5568	.4970	.3971	.3186	.2567	.2076	.1685	.1372	.1122	.0920	.0687
13	1.0000	.7730	.6006	.5303	.4688	.3677	.2897	.2292	.1821	.1452	.1163	.0935	.0754	.0550
14	1.0000	.7579	.5775	.5051	.4423	.3405	.2633	.2046	.1597	.1252	.0985	.0779	.0618	.0440
15	1.0000	.7430	.5553	.4810	.4173	.3152	.2394	.1827	.1401	.1079	.0835	.0649	.0507	.0352
16	1.0000	.7284	.5339	.4581	.3936	.2919	.2176	.1631	.1229	.0930	.0708	.0541	.0415	.0281
17	1.0000	.7142	.5134	.4363	.3714	.2703	.1978	.1456	.1078	.0802	.0600	.0451	.0340	.0225
18	1.0000	.7002	.4936	.4155	.3503	.2502	.1799	.1300	.0946	.0691	.0508	.0376	.0279	.0180
19	1.0000	.6864	.4746	.3957	.3305	.2317	.1635	.1161	.0829	.0596	.0431	.0313	.0229	.0144
20	1.0000	.6730	.4564	.3769	.3118	.2145	.1486	.1037	.0728	.0514	.0365	.0261	.0187	.0115
21	1.0000	.6598	.4388	.3589	.2942	.1987	.1351	.0926	.0638	.0443	.0309	.0217	.0154	.0092
22	1.0000	.6468	.4220	.3418	.2775	.1839	.1228	.0826	.0560	.0382	.0262	.0181	.0126	.0074
23	1.0000	.6342	.4057	.3256	.2618	.1703	.1117	.0738	.0491	.0329	.0222	.0151	.0103	.0059
24	1.0000	.6217	.3901	.3101	.2470	.1577	.1015	.0659	.0431	.0284	.0188	.0126	.0085	.0047
25	1.0000	.6095	.3751	.2953	.2330	.1460	.0923	.0588	.0378	.0245	.0160	.0105	.0069	.0038

Kiona International

Table 6-1. Discount Table

An *investment grade* audit is a basic ingredient of a strong bankable project. An audit that focuses on investment guidance speaks the financier's language and makes projects even more attractive. Done right, the process will make more financial options available and bring down interest rates.

An IGA is more costly than a traditional audit, but these additional costs can easily be offset by lower interest rates as well as the ease of obtaining financing.

FINDING THE MONEY TO IMPLEMENT AN IGA

One of an auditor's greatest frustrations is to have performed a quality audit showing significant potential savings and then have an owner do nothing with it. The most frequent reason cited by owners for allowing audits to gather dust is the lack of money. Either the client's organization is strapped for funds, or other needs within the organization seem to take precedence.

Some auditors don't view financial guidance as within their purview. Auditors, however, who are inclined to perform investment grade audits, seem to exhibit a much broader sense of responsibilities to their clients—a responsibility that often includes implementation support and the identification of funding sources. To meet the needs of these more conscientious auditors, who are anxious to see their recommendations bear fruit, it may help to explore the implications of getting IGA recommendations financed.

First, let's look at some other reasons that energy efficiency financing resistance raises its ugly head. Frequently, the lack of action can be traced to the inability of business office personnel to understand "energy" needs and potential savings. All too often, they view such "boiler room" talk with trepidation.

Most owners don't buy "kWh's"; they buy lighting, motor power, conditioned air, etc. For the owner, then, the total cost of energy can be measured in the useful work it performs. Management regards the energy required to cost-effectively increase production in an entirely different way than the kWh's required to light a new sign or put a light bulb in that corner bathroom in the warehouse. To get management's attention, energy calculations need to address useful value in relation to the amount of energy saved and its associated costs—and ultimately, the net financial benefit to the company.

For the auditor, who is on a "money mission" for an industrial client, two frequent mental barriers present themselves: 1) self-imposed return on investment (ROI) parameters where the payback is limited to 2 years or less—too often passing up a 50+% ROI while money sits in a equipment reserve or replacement account earning less than 6%; and/or 2) investments in new production lines have more appeal than reducing operating costs. The first barrier fails to recognize that no new budgeting is required—it's money now going for wasted energy. The second reason really misses the mark because reduced operating costs can make a company more competitive plus the net benefit from an energy efficiency investment generally requires about one-fourth the investment to attain the same financial benefit as selling more product.

Try it. Take the numbers from the IGA and calculate what the net financial benefit (savings minus investment) for the life of the project.

Then, calculate what would have to be invested to get this same value in net profit from increased product sales. Keeping in mind that after incurring all these production expenses, there is a very real possibility that not all of this additional production will be sold.

Delaying work is a management option. Top management, however, needs to recognize that putting off energy work is not the same as delaying a painting job. Painting the building can be delayed until next year with only slightly elevated costs, but delaying energy work creates a permanent loss. Once the wasted energy dollars go up the smokestack, they are irrevocably lost. You can get the same effect by piling a bunch of money on the CEO's desk and putting a match to it.

"SURE ONES COULD SAVE ALOT, BUT 50s FEEL SO MUCH WARMER"

As the illustration suggests, sometimes management knows they are burning money, but the wrong solutions sometimes get in the way of solid audit recommendations.

A Buyer's Market

If an auditor wishes to go the next step and help the client get the most favorable loan, it's good to remember the financial houses need you and your client more than you need them. Too often, people approach lenders as supplicants. A "hat-in-hand" approach is 180 degrees out of phase with reality. For banks or any financial house, money is the commodity that they sell—well, more realistically, it is what they "rent." To stay in business, they have to have someone to rent it to. When you present them with a quality project to fund, you are doing them a favor.

Financing energy efficiency is definitely a buyers' market. In most instances, the sources out number the viable projects to be funded in North America; so it pays to shop around.

Going Shopping

Comparing lenders can be a very profitable piece of business. Beyond the obvious interest rates and financing charges, there are a number of other factors that should be carefully assessed. Once an auditor starts evaluating financial sources for a client, the criteria offered below seem to fall into place and can be adapted to meet each client's unique needs.

Since it's a buyer's market, encourage your clients to look for the following qualifications in the financing sources they are considering. Or, you or your firm may find qualifying lenders to be a service that your clients will value—and pay for.

Experience

When projects are viewed from the financier's point of view, it becomes readily apparent that interest rates are predicated on perceived risks. If a banker is not familiar with energy efficiency (EE) financing and not comfortable with doing the necessary EE project due diligence, they are apt to see more and greater risks. When they perceive greater risks (whether they are real or not) they will assign higher interest rates.

If, when an energy savings project is presented, bankers lapse into talking about collateral and balance sheets, it's usually a dead give away that they are outside their comfort zone. It is reminiscent of the banker we talked with in Moscow, Russia, who told us, "We don't loan money for hot air!"

Another indicator is the presence of an engineer on staff, or a financial house that regularly retains one. If no engineer is available, it begs the question: Who is going to evaluate the technical aspects of the project? If not evaluated correctly, the bank is taking on greater risks and… the interest rates go up some more. While not quite as critical, the same concerns can be raised for having in-house legal, tax, and accounting expertise.

Another area well worth checking is the typical size, or range of project sizes, the bank normally funds. Also, it might help to identify which market segments, if any, the financial house typically focuses upon. For example, some lenders specialize in municipal leases and understand tax-exempt implications for a certain client base. If your client is a government entity, *and qualifies for tax-exempt status,*[6-3] this may be a critical consideration.

Bottom line: Lenders with experience in EE financing are apt to process a loan more quickly and set lower interest rates.

Flexibility

Over the life of a project, the flexibility of the financial house related to documents and payment structures can affect the profitability of a project. Their willingness to customize the loan package, such as accommodating the client's budgeting process, can make a tremendous difference.

Flexibility paves the way for finding unique ways to leverage, syndicate, or secure the financing.

The basic documents may be standard, but flexibility allows for meeting the client's diverse needs. Some institutions act like the documents are cast in bronze and they cannot deviate from the printed word. It pays to remember: In the wonderful world of word processing, flexibility is always possible.

Level of Service

Imbedded in the above criteria is a very basic theme—the inclination of the financial house to serve the client.

When the Hansen contingent moved "home" to the Pacific Northwest a few years ago, inquiries were made about banks. We were told that the head of Columbia Bank had left another banking system because it did not offer the customer a level of service that he thought it should. That sounded good, so we sought out our local branch. We didn't know banks could, and would, still give this caliber of service! For instance, the difference between our former Maryland bank and Columbia in tracking wired funds from other countries and keeping us informed was, and is, incredible. Not only is the service extensive, it is friendly and cheerful.

The moral of the story; it still pays to shop around for yourself and/or your client. As you walk through their door, keep in mind that their business is renting money and serving you—and they need you to make it work.

Transaction Costs

To process the papers and do the due diligence on a project represents a bank's operating costs, typically referred to as transaction costs. The amount of the operating, or transaction costs, in relation to the amount of a loan sets a level of business efficiency. In fact, banks keep track of, and report, their "efficiency ratio," which is based on the cost of earning $1 of revenue. Since it takes almost as much time and effort to process a small loan as a big one, it quickly becomes clear why financial people prefer larger projects.

Acting on such information, some auditors, or the firms they work with, have found additional business in aggregating several small projects to receive more favorable financing terms. Pool financing not only creates a bigger "package," and, therefore, more favorable financing; but it has the added benefit of spreading the risks over more projects. And it just keeps getting better since spreading the risks, in turn, enhances the financial terms.

IGAs FOR ESPs AND ESCOs

If energy service providers (ESPs) and energy service companies (ESCOs) are involved in the process, it can have a major impact on the audit and on the project.

First, let's differentiate between an ESP and an ESCO. Both offer essentially the same energy services. The major distinction is that an ESCO makes guarantees, usually guaranteeing that the energy savings, with certain caveats, will cover the cost of the project.

The way an ESCO structures the deal financially will also affect the audit. The two dominant world performance contracting models, shared savings and guaranteed savings, have common auditing needs, but also impose separate conditions. For example, shared savings is based on a predetermined percentage split of the energy *cost* savings, while guaranteed savings relies on guaranteeing the amount of *energy* to be saved. Since shared savings rests on future energy prices, the risks are much greater. When energy prices fall significantly, as they did in 1986, projects and even ESCOs are put in jeopardy. Bankers have long memories and recall the defaults in the late 1980s.

Guaranteed Savings

Of considerable interest to the IGA auditor, conditions related to ESCO guarantees can:

a)   **Limit options**. Guarantees force the establishment of parameters as to what equipment can be used within the guaranteed effort; i.e., only equipment with a proven track record for savings as well as a record of the costs for parts and maintenance can be recommended if the ESCO is to guarantee its performance. (If an owner wants new technologies or equipment, which do not have the necessary track record, they can still be recommended, but outside the guarantee.)

b)   **Require certain conditions**. Guarantees necessitate more control. Specifically, ESCOs want more control over matters related to installation and maintenance provisions. (ESCOs are very aware of the study, reported earlier, that up to 80 percent of the savings in an effective energy management program can come from the energy efficient practices of the operations and maintenance [O&M] personnel; so leaving O&M to chance is pretty scary stuff for an ESCO.)

c)   **Lower interest rates**. On the positive side, guaranteed savings can reduce the interest rates financiers charge as risks are more clearly identified, assigned, managed and/or mitigated in a risk-based project, such as performance contracting.

Shared Savings

A riskier, and therefore more costly model, is shared savings. In addition to the uncertainty related to energy pricing, the following aspects of shared savings suggest why this option has a major impact on project financing.

a) **Hedging the bet**. If you consider how paybacks are determined:

Payback = amount (energy saved) × price (of energy),

there must be protection from a price drop risk. To preserve the payback (or the dollar value of the project), ESCOs must balance this risk by increasing the amount to be saved or their percentage of the split. In either case, less of the investment will go into the project and into potential savings as more must be dedicated to covering the risks. When the ESCO shared savings risk cushion is paired with the risks perceived by the bankers, it is not unusual for a shared savings project to be about one-third smaller than a guaranteed savings project for the same level of investment. These limitations seriously impact the range of measures that can be considered in an audit.

b) **Shorter contracts**. Given the threat of price volatility, it is not surprising that ESCOs tend to have projects of shorter duration if shared savings is involved. A shorter contract forces a shorter payback and constrains the auditor as to the measures that can be considered.

Rather dramatically, the economic viability of an entire project can disappear if in addition to the savings limitations, other concerns are identified. Such concerns include problems, such as initial low energy prices, or measures with relatively long paybacks.

c) **Carrying the credit risk**. Another financial implication rooted in shared savings, which may not be readily apparent to auditors, could be critical if the client is an ESCO. In shared savings, the ESCO typically carries the credit risk. If the ESCO finances just one or two projects out of many, it may not be a problem. If, however, the ESCO finances several projects and does not belong to a large company, it can quickly become too highly leveraged to take on any more debt. An ESCO in such a position generally has three options: 1) sell the paper (Once the project has been installed, the contract is sold.); 2) carry on a parallel business to keep things going; or 3) limit projects to very short paybacks to maintain the cash flow. If the third provision is put into play, a quality investment grade audit would tend to be over-kill. One final rather important thought, ESCOs, which are heavily involved in shared savings can be financially strapped; so the auditor might want to take precautions to be sure he or she gets paid.

Given all the disadvantages associated with shared savings, it is a wonder that it is used at all. However, it is very attractive in situations where the customer is stable, but is cash poor and/ or has limited creditworthiness. This is particularly true in the case of weak or transition economies, which are often found in developing countries. In other instances, a shared savings project can be structured off balance sheet, which can be very attractive to an owner who's organization is bumping against its debt ceiling, or who fears what additional debit will do to the organization's credit rating. Others, such as the US federal government, are so risk adverse that they are willing to suffer much greater loss of investment potential rather than incur the credit risk involved.

The reference in the previous paragraph to "off balance sheet" brings up another factor that indirectly impacts the audit when shared savings is involved—the whole question of capital leases vs. operating leases—since operating leases are often considered in off balance sheet financing. Rigid provisions exist as to what constitutes a capital lease. If a lease meets ANY of the following criteria, it is considered a *capital lease*:

•   the lease term meets or exceeds 75 percent of the equipment's economic life;

- the purchase option is less than fair market value;

- ownership (or title) of the equipment is transferred to the customer (lessee) by the end of the lease term; or

- the present value of the lease payments is equal to 90 percent or more of the fair market value of the equipment.

Conversely, if a leasing arrangement meets any of the above criteria, it cannot be an *operating lease*.

For the sake of comparing the finance models' impact on audit procedures, guaranteed savings and shared savings have been put at opposite ends of a continuum. In practice, it is quite possible to move either model toward the center and create hybrids. Audit procedures would then need to be adjusted accordingly.

INTEGRATED SOLUTIONS, OR CHAUFFAGE

There is an entire range of financial issues related to supply options. Auditors have historically dealt with demand side issues, but integrated solutions (supply and demand services provided by the same prime contractor), also known as chauffage, may require an auditor to broaden his or her repertoire. This might include new expertise related to contract valuation and pricing, valuation of real assets, derivative asset valuation strategy, forecasts and more. If faced with such a need, an auditor might seek specialized training such as that offered by the Electric Power Research Institute (EPRI), or some type of an association might be created with an energy professional who has a command of supply-related financing issues.

Whether an auditor seeks to achieve some level of competency in this area, or chooses to partner with someone who has such competency, it pays to be aware that managers in the energy and power industry have new worries to go along with market price and load volatility—vanishing liquidity and stringent credit rating standards are becoming dominant themes. In the post-Enron era, energy and power companies must establish comprehensive risk management practices and procedures. In addition, they now must make their businesses more transparent to creditors and investors. Such requirements can work to the benefit of the consultant looking out for a client's supply interests.

If the reader wishes to get a better idea of the financing mechanisms available when serving the ESCO industry, earlier books by Shirley Hansen, *Performance Contracting: Expanding Horizons*, discusses these mechanisms in lay terms for the non-accountant, and *Manual for Intelligent Energy Services* provides the same material from an owner's perspective.[6-4]

CREATING A BANKABLE PROJECT

What is a "bankable" project? Simply put, it is a clearly documented economically viable project.

In recent years in North America, financing has been so plentiful that it can be said, "If a project cannot be financed, it's time to rethink the project." Creating a bankable project in North America has, therefore, come to mean putting together a project proposal that will receive the most favorable treatment from the lending houses.

In countries outside of North America and parts of Europe, creating a bankable project is much more demanding. This is partially due to the fact that bankers are not fully acquainted with energy efficiency financing procedures, and partly due to proposers who are still not clear as to what bankers are looking for.

Whether creating a bankable project focuses on getting the funding or getting better terms, **an investment grade audit is at the heart of a bankable project**. The identification of project risks plus the costing out of potential management/mitigating strategies really facilitates the due diligence process and is music to bankers' ears.

Donor agencies, such as the World Bank, European Bank of Reconstruction and Development, and the US Agency for International Development, have helped banks and proposers in developing countries improve EE financing procedures. They have provided funds to cushion the risks for the banks, established guarantee funds to aid banks in dealing with financing uncertainties, and offered training to bankers and proposers in energy efficiency financing requirements.

The donor agencies, however, can only take the process so far. Inevitably, if an energy efficiency industry is to be established in a country, the local commercial banks and financial houses must become active in financing such activities.

In all instances, the investment grade audit facilitates the process. Financiers need to be made aware of what an IGA offers and how the related benefits are achieved.

Creating a bankable project starts with sorting out the pieces that make a project economically viable. As stated in the previously cited book about performance contracting, "The first step is to examine the key components and make sure each aspect is properly assessed and the plan to effectively manage that aspect is clearly presented. Each component carries a risk factor... and each risk factor carries a price tag. An effective ESCO knows how to assess the components and how to package them into a project that can be financed."[6-5]

Whether or not an ESCO is involved, the identified components for seeking funding are essentially the same and include a pre-qualification of the customer, audit quality, equipment selection and installation, project management, and savings verification. In addition to this, the bank will want to have the necessary information about the customer, ESP, engineering firm, and/or the ESCO to establish their creditworthiness. The reputation and history of the borrower can often add surety, which in turn offers financiers added confidence. "Added confidence" can be translated into lower interest rates.

Creating a bankable project, therefore, has two key aspects: 1) to get the project financed; and 2) to get the lowest possible interest rates. Firms that regularly seek energy efficiency financing understand the critical importance of a quality investment grade audit in developing a bankable project and getting the most favorable terms.

References
6-1   For more information, contact Dr. Richard B. Jones at 18 Pale Dawn Place, The Woodlands, Texas 77381.

6-2   The discount table presented on page 85 is taken from Kiona International's finance curriculum materials. The formula at the top of the table provides the means to derive the figures presented in the table. Kiona International can be contacted through Dr. Shirley J. Hansen.

6-3   The reader should be aware that just because an institution does not pay sales tax, it is not necessarily tax-exempt. Section 103a of the Internal Revenue Code sets forth the tax-exempt conditions.

6-4   Hansen, Shirley J., *Manual for Intelligent Energy Services*, 2002. pp. 98-103. Hansen, Shirley J. and Jeannie C. Weisman, *Performance Contracting: Expanding Horizons*. 1998. pp. 76-82. Both books published by The Fairmont Press, Lilburn Georgia.

6-5   *ibid.*

Chapter 7

Working Risk into the Mix

I
T IS WORTH SAYING at least one more time: From a basic technical standpoint, there is nothing wrong with the traditional energy audit. All of the data that has been gathered in the past is still necessary. The difference between what is needed today and what was acceptable yesterday is the identification and financial quantification of risk.

This is a good place for a short review of some terms we have been using with great frequency but perhaps with insufficient definition. First, what do we really mean when we say "traditional audit?" In very broad terms, the traditional audit seeks to discover energy waste and determine measures to minimize that waste. This audit will normally determine costs for the measures recommended and provide calculations of expected paybacks. This is a "snap shot" approach, essentially assuming that conditions, at the time of the audit, will not change over time.

The Investment Grade Audit (IGA), in equally broad terms, seeks to discover and minimize waste while simultaneously considering, discovering and minimizing investment risk. The IGA looks at what is happening now and then reaches ahead to consider what might occur over the life of the project. This audit includes the cost considerations attendant to the measures to be installed and goes beyond that to look at the possible risks that may impact the financial outcome of the project. The IGA will include an inventory of those risks, an evaluation of their potential impact, an analysis of the ways these risks may be managed or mitigated and the attendant costs. These costs, having a direct impact on the financial viability of a project, would rarely be considered under the terms of the traditional audit. While the traditional audit is a snap shot, an IGA might be considered more like time lapse photography with the opportunity to examine what happens (or may happen) over time.

It is easy to recognize the progression from the traditional audit to the much more sophisticated and complete investment grade audit, but the progression will not stop there. The final chapter of this book discusses Master Planning, incorporating the approaches of an IGA into a broad based document providing a dynamic roadmap for the intelligent management of a facility. It provides critical guidance in ways change can be anticipated, planned for and budgeted in an orderly manner.

Traditionally, engineers have lived in a Pass/Fail world; a world where answers are either right or wrong… savings better than estimates (Pass), or saving less than the estimates (Fail). Formulas have not been dependent upon truth; truth has been dependent upon formulas! Under those conditions, variables within those formulas must be straight forward and predictable.

Today's real world operates somewhat differently… comfortable predictability is getting harder to find. Many things that we now know can make the difference between the success or failure of an energy efficiency project were never mentioned in an engineering handbook. We have discovered that opinions have impact on the outcome of efficiency projects. A wide variety of opinions regarding comfortable room temperatures, lighting levels and color rendition, the value and use of windows, the amount of maintenance required and a host of other opinions. Opinions may be slightly changed through application of reason and logic, but almost never fully changed.

We have also come to accept the fact that buildings and industrial processes are never static. They present a moving target. What seems set in stone today as far as building occupancy or function, or the activity in a manufacturing plant, can look very different a year from now. A school can add a large computer laboratory, open a gymnasium for community use, take in more, or fewer, students… all things that may make significant differences in energy usage. A manufacturing plant may change product lines, vary production levels, add or eliminate shifts… changing energy consumption or use patterns. These things and a myriad of others, present risks; risks that here-to-fore were often overlooked or understated. But if they are recognized and considered at the inception of a project, they can be quantified and, usually, managed.

Engineers have always assessed risks, but principally those risks related to the accuracy of their work and/or the practical applicability of the measures they were considering. They considered what impact of the measures they proposed would have on other equipment or functions within a host facility and they took into account their assessment of the ability and skills of those who might operate the equipment they would install. These are real risks and, at times, the basis of "go or no go" decisions. It might be said that the traditional audit took us from "A" to "T." Now we are reaching for "Z," a level where risks are assessed in much broader terms.

THE NATURE OF RISK

The idea of formalized risk assessment, mitigation or management may seem a little foreign at first glance, but it is merely taking something we all do every day, usually without conscious thought, and applying it to business problems. From the time we awaken in the morning until we go to bed at night (and sometimes even there), we assess a multitude of risks and deal with them, assigning values and applying mitigation strategies. Consider "jaywalking." It presents a risk we all recognize.

The down side of jaywalking, the potential "cost" is rather obvious… being hit by a car or getting a ticket. Of course, if our risk assessment is not complete, there may be risks, such as flying pianos, that we fail to consider. The possible benefit of jaywalking is the time and effort saved by not going to a crosswalk. This would also be the cost of mitigation. But how about "managing"

the risk? Careful observation and evaluation of the traffic could be a management strategy; there is still risk, but it has (hopefully) been reduced to an acceptable level. We don't spend a lot of time thinking this sort of scenario through, but that is the procedure we follow. Again, usually without conscious thought.

Risk in Financial Transactions

Risk guides the thinking in every financial transaction. "Am I getting my money's worth? If I invest in this project, will I get my money back with a profit, or will I take a loss?"

As discussed in the financing chapter, financial institutions are in the business of "renting" money. They make loans to people whom they hope will use the money to make more. The financier will look at the loan applicant and (like a gambler) consider the odds of the loan being repaid. His or her evaluation of those odds set the interest rates. Interest on a loan (the rent) is the sum of the cost of the money to the bank (what they pay for it) plus a profit, plus a risk "cushion" based upon their evaluation of the likelihood of the loan being repaid.

Due diligence, the bankers name for risk assessment, is the process they go through to decide whether they will fund a project and if they will, what rate of interest will be charged on the loan. One of the most important functions of an IGA is to assure a lender contemplating the financing of an energy efficiency project that you know what the risks are and have taken them into account in designing the project. Your "due diligence" can provide the lender a confidence level that can make a real difference in the cost of money over the life of the project.

Risk Assessment

In simple terms, the risk assessment we must do for an energy efficiency project, whether an "in house" effort, a project assigned to an energy engineer, or a comprehensive project to be accomplished by an ESCO, comes down to a few basic steps.

Step 1 is a realistic inventory of the risks likely to be involved, especially as they relate to the energy saving measures under consideration. The matter of time becomes particularly important where guarantees are involved. It gains importance when payments are based upon the efficiencies obtained over a period of years, as is done under a performance contract.

Step 2 is determining what risks are manageable. How, through the design of the project, can those risks be handled without damage to the overall outcome? Can they be accepted and dealt with?

Step 3 assesses which risks can be, and should be, eliminated from the equation. If there is a way to remove the risk through planning or engineering this will become a cost factor to a greater or lesser degree. Those costs may determine the financial viability of a specific measure, or even the entire project. The second and third steps often overlap and may not be sequential.

Step 4 involves an assessment of potential mitigating strategies and their associated costs. Risks are often assigned to suppliers. For example, lamp suppliers can be asked to guarantee the mortality rate on their lamps. Manufacturer warrantees can be extended to include other equipment performance matters, or energy savings characteristics.

If outside contractors are involved, especially in the case where a performance contract is to be utilized, the assignment of risks as a mitigating strategy becomes important. Who bears the cost if systems do not perform as designed over the life of a project? Who gets hurt, who bears the burden, if there are changes in the function of a facility affecting the costs or savings of the efficiency project?

In performance contracts, risks are usually (and best) assigned to the party in the best position to manage them.

☐The manner in which the above steps are implemented will determine the success of an energy efficiency project, not only in terms of energy saved, but in monetary value to the owner and contractor. A careful assessment of risks, for example, allows an ESCO to narrow the "risk cushion" normally used to account for the unknowns likely to be faced in a project and make it possible to guarantee a larger percentage of the calculated energy savings. This permits larger projects, often at the same investment level, which can increase the benefits to the facility owner and ESCO alike.

In general, the risks involved in energy efficiency projects fall into two broad categories: people factors and facility/process dynamics. These factors have little meaning if the "project" consists of a single measure, or piece of equipment has been bought on a bid basis with limited warrantees and no guarantees of energy saving performance. But, if that piece of equipment is to produce specified savings over time, or if savings are guaranteed, all the risk factors apply. When money rides on project outcome, the full appreciation of risk by all parties becomes vitally important.

PERFORMANCE CONTRACTING

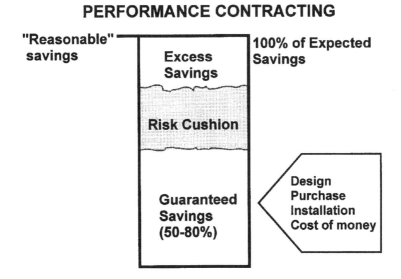

Figure 7-1. Managing Risks through Financial Structure

FITTING THE PIECES TOGETHER

When we move beyond defining what an IGA is and assess how well the pieces fit together, it is clear the results benefit facility owners, engineers and contractors. A well-done investment grade audit benefits everyone involved.

□The approach to risk management in an energy efficiency project can be expressed as a series of quite logical steps leading to the realistic calculations of the payback periods for the measures you may want to include. The illustration below outlines the steps for each measure and you will note the arrow is labeled "The reality check." This is most appropriate because the process outlined is designed to determine and deal with what is real; not just what an ideal situation would be for each measure contemplated.

Risk Inventory

Once a potential measure is decided upon, the first of the steps listed is "identify risks." This was referred to earlier, quite appropriately, as an inventory. At the outset it is simply a list of things that may hamper the success of a given measure (and ultimately a project), ranging from actions of people to equipment problems. This often can begin as a checklist, created off-site, that

Figure 7-2. Determining the Real Costs of EEMS

includes a number of items collected from experience. It will include such things as management's attitude, skill levels of O&M personnel, personnel turnover, type of facility occupants (school, office, industrial, etc.), and future plans for the facility. For an industrial installation, contemplated changes in product mix or hours of operation must also be identified. The list can become rather long but it will serve as a valuable guide when the actual on-site examination begins.

Once on site, items will be added or removed as appropriate, but the key is to be inclusive... save the "weeding out" for the next step. A little pessimism when looking at things that may present a risk, can pay off in a big way. A time for optimism will come later if you have done a good job of anticipating all the rough spots.

The People Factor

Keep this thought: *"Find the person who can kill your project... he may be the person who can make it work."*

We have come about as close to predicting technical reality as we can through our engineering calculations. Engineers must now face their greatest challenge... we must begin to incorporate people into our predictions! Not what people should think, or how they *should* act, but how they **do think** and **do act**.

If we ever hope to produce audits closely resembling reality, we must accept the fact that energy is consumed by people, but more importantly, for people. And it is those people who will ultimately make, or break, an energy efficiency program. Since the "people factor" is probably the least understood facet of the investment grade audit, and the source of most of the risks in any project, it is a good place to stop and take a good look at the occupants of the facility... all the occupants. In Chapter 3, we highlighted the need to weigh human behavior and stressed that what the people in a facility believe, disbelieve, do, or do not do, can spell the difference between resounding success and costly failure. Now we need to take a closer look at the risks these people represent and how that risk potential might be assessed. A checklist that considers some major concerns follows. In practice, items will be added or deleted to fit certain clients.

☑ TOP MANAGEMENT

☒ *Commitment*
The attitude and actions of the top management of a facility, as discussed in Chapter 3, can make all the difference in the outcome of an efficiency project. As noted earlier, one engineering firm places so much emphasis on management commitment that it will not even begin an audit of a facility unless a top management policy statement embracing the idea of energy efficiency is in place. The owner of the firm is convinced that unless a policy is in place, and has been communicated throughout an organization, a high level of efficiency just will not happen

☒ *Policy*
This "policy" must be more than just an "efficiency is good" statement. It must recognize that energy efficiency improvements are important to the success of the organization and demonstrate a recognition that energy efficiency relates directly to the bottom line. Along with a general policy, it is important that responsibility for energy efficiency be assigned and recognized as significant on the "report card" of management personnel.

At this point it might be sufficient to say, "If the top brass doesn't care about energy efficiency, it is highly unlikely that any project will be very successful."

When you begin to evaluate an actual facility, a critical first step is to get a solid feel for the attitude of the top management. If a real energy policy is in place and if the various levels within the organization have "bought in," things are looking up. But it doesn't always work that way.

☒ *Occupants*
The occupants of a facility often present some risks that are well worth considering. How much control will occupants have over

conditions in the facility? We can be sure the temperature that satisfies the teacher in one schoolroom will be far too hot or much too cold for another teacher down the hall. Although seemingly trivial, how the building custodian solves this problem can present a risk, and that risk must be accounted for. Of course a major occupancy risk involves changes in numbers of people in the facility or changes in the nature of their activities. Since occupants are not totally predictable, the list grows.

☒ *O&M*

Top management can make policy, but it is far down in an organization where you find the people who make things happen. Or not happen! It is at the operations and maintenance (O&M) level where risk assessment becomes more complex and where mitigation and management strategies play a larger role. It could well be carved in stone: "If the director of operations or the maintenance manager do not believe a project will work, **it will not work**." That is a risk that requires a prominent spot on your list.

If a top down efficiency policy is in place and widely understood perhaps half the battle of risk assessment and mitigation/
management at the O&M level is already won... but not necessarily. What about the skill levels and capability of those who actually operate the energy using systems on an everyday basis... can
they deal with the more sophisticated systems your audit suggests
would be ideal? Are adequate maintenance procedures, including
preventive maintenance in place *and in use*? And is the rate of
personnel turnover such that continuous training would be
needed? More risk items to be considered.

☒   *Past Practice*

An Investment Grade Audit must include some level of analysis
of past maintenance practices, recent repair costs for problem
equipment, preventive maintenance techniques, and staff skills.
When the Maintenance Department is not provided the training,
tools or staff to keep up with the continuing growth in the technical level of equipment being added, problems arise... not just
equipment problems, but attitude problems.

☒   *Housekeeping*

A pretty good indicator of attitude can come from the simple
observation of the "housekeeping" in equipment spaces and
around machinery. Poor housekeeping often indicates something
less than adequate maintenance and also may suggest that someone in the O&M chain of command lacks pride in his or her work,
or just thinks nobody cares anymore. A risk that could have serious impact unless it is dealt with.

☒   *Maintenance Schedules*

An important indicator that may, or may not, be related to morale
and/or attitude is the presence, or more importantly the absence,
of two written schedules; a maintenance schedule *and* a preventive maintenance schedule for every piece of equipment in the
facility. If these schedules are in place and in use, risk levels are
reduced even with the existence of someone who is less than convinced that an efficiency project will work. If they are *not* in place

and *in use*, allowance will have to be made for the establishment of these programs before an energy efficiency effort can realize its full potential. The cost associated with establishing a solid maintenance program will have to be factored into project costs if contemplated system changes and upgrades are to be successful. The benefits will far surpass the costs over time.

☒ *Needs and Attitudes*

Auditors need to be aware of the needs and attitudes within the Maintenance Department *before* expensive new systems are installed. This concept has already become generally accepted among third party investment organizations, but their remedy-of-choice is all too often wrong. As noted in Chapter 3, their tendency has been to charge more money to a project for additional outside maintenance on the equipment being installed. This leads to an unrealistic division of responsibility within a maintenance department. "Whose job is it?" separates the in-house staff from any "ownership" of the efficiency program. As previously noted, some projects will require a highly skilled technician for proper maintenance, but those projects are rare. It is also true that there

can be an in-house tendency to use added staff for services which will draw away from servicing equipment specific to the efficiency project. In most instances, these issues can be handled through the administration and, in fact, may even prove to be an advantage to the project.

An improvement to the capabilities of a maintenance department can go beyond assistance tied to pieces of equipment specific to the energy efficiency project. The efficiency "investment" can take on a lasting impact within the maintenance group by improving skills while simultaneously providing full time guardianship over the key newly installed systems.

☒ O&M Skills

The skill levels of operations and maintenance personnel present another area of risk assessment where values and costs must be considered. Over time, most of the energy savings in an energy efficiency project come from the continuing efforts of people, it is imperative that those, who do the maintenance, push the buttons and operate the machinery, be capable of doing their jobs right. The inherent risks are obvious. If those O&M people are not fully capable of maintaining and operating the equipment and systems key to an energy efficiency program, it will not produce the savings that could have been achieved.

☒ Training Needs

After having established that skill levels are good and maintenance programs are in use, what about personnel turnover? If the turnover level is significant, many of those skilled people may be gone well before the end of your project. A serious risk. The cost of mitigation will include the training necessary to assure the continued availability of skilled people to operate and maintain the key systems, which in turn will determine the results of the energy efficiency effort.

It would be a mistake to leave the discussion of an operations and maintenance function without acknowledging the unique role the O&M people play. They frequently are the folks who know the most about the overall operation of the facility and have a feel for the peculiar needs and demands of the people and systems than make up the whole. They are an important resource as well as a potential stumbling block and an auditor overlooks their value at his or her peril.

To emphasize this point, remember that it is useful to understand why systems of concern are operated as they are. Sometimes the reasons may be obscure, but the results pretty dramatic. Here is a real life example from Energy Systems Associates.

> *We were ready to install a new control system when we discovered that the central chiller was left on around the clock, seven days per week. Why? Because the building was built over an underground spring and the floor tile oozed glue up through the cracks whenever the room humidity level got too high. The only person who knew this was the head custodian. So much for that $45,000 "energy saving" controls investment.*

☑ THE FACILITY

An essential truth to remember when weighing the risks—a truth, which is apt to have a major impact on the results: A building, a factory, a system is a moving target. Changes will occur! We can add a line to the old Casey Stengel quote, "Life ain't what it used to be... " and say, "it ain't going to be like that next week either."

What about the future of the facility in question? We can never be positive but we need to find out, as best we can, what the long-term plans for the facility might be. Are they expanding activities, adding a new wing, changing from a school function to administrative offices? Are they likely to go out of business? Real risks! We know of a fairly long term energy performance contract written on a school building that was *scheduled* to be closed long before the end of the contract. Not much risk assessment there, but an interesting court case.

The situation for an industrial facility is not much different. Contemplated changes in product mix can change energy use patterns. Adding or dropping a shift can make a real difference in the viability of a project. Substantial changes in personnel numbers can also make a difference. And it pays to wonder, "Is what this plant produces something that will be around for the long haul?" A modern, well equipped, plant that is still producing figurative buggy whips may have to go through major changes soon, which could make a real difference to your project. Production changes very frequently mean changes in energy demand... increased loads or decreased loads; changes that can impact the financial results of a project.

☑ TECHNICAL RISKS

There are, of course, other risks that must be considered when evaluating a potential project; some of a technical nature, which were sometimes neglected in the traditional audit because their impact was not considered significant in the overall package. Today, with more attention being paid to narrowing the gap between estimates and outcome, they are well worth further consideration, particularly with regard to the interaction between people and technology. Auditors must be constantly up-to-date to be aware of, and understand, the significance of new technology as it enters the market. With the ever increasing impact of technology and the growing tendency of facility owners to look for the newest and most "high tech" solutions, it becomes important to look beyond the major energy using systems to examine the sub-systems that may have an effect on rates or timing of energy use.

☒ High Tech

The attraction of "high tech" can itself pose risks that can be significant, especially where performance guarantees come into play. Everyone wants to feel they are "up-to-date" and ready with the very latest ideas and equipment, but this can easily become a trap. What is needed, when guarantees are in place, is the application of proven systems and equipment that have shown over time they will produce the efficiency levels required under the specific conditions of the project. Installation of the latest "black box" may impress a facility owner, but if the results are guaranteed over an extended period the "very latest" may be a gamble not worth taking.

☒ *Low Tech and Common Sense*

Beyond high tech considerations, there are some quite low tech risks that have been known to bite. A classic case focuses on some engineers, who attempted to apply computer-based controls to duty cycle a 100-year old boiler. The boiler worked fairly well until the controls took over, then it died. In this case "the twain" never met.

☒ *System Compatibility*

Compatibility of equipment or systems is a frequent problem with results ranging from near disaster to less than optimum efficiency improvements. It frequently pays to back away from the best of the high tech solutions for a little less sophisticated approach, which will mesh well with systems already in place and match the capability of the people who operate them. Increasingly, open protocols and software, which open these doors can help, but the auditor must remain cognizant of the need.

☒ *Ancillary Equipment*

Ancillary equipment can often have a real impact on a retrofit project. What looks right can sometimes turn into a technical problem all by itself. Here is what one of Energy Systems Associates' engineers ran into on a job:

> We replaced an old 300 ton chiller with two new 150 ton machines—only to discover that the pneumatic 3-way valve controlling flow to the cooling towers had the tubing connected backwards. It had been that way for more than 20 years, but it just hadn't mattered to the old chiller. As you might expect, our little 150 ton units had some trouble with that!

Careful evaluation of systems as they are, before changes are made, can avoid some headaches as well as possibly expensive risks.

You can look at the way energy is used within a facility as a chain, with links connecting the end point, a light bulb, a heat vent or a specific piece of machinery back to the meters where electricity or gas enters the system. Taking a careful look at the energy "chain" from the far end, can help find potential problems hidden as facilities have been altered or equipment has been added over the years. Start where the lights are burning, a machine is running, or heat is coming out of a vent; then, follow the chain back to the meters. Are there links that no longer serve a purpose beyond being a drag on the system? Frequently, the energy chain gets stretched when something new is added with no thought of what could, or should, be dropped. Too often, this leads to overloaded or badly organized electrical circuits, ductwork or compressed air lines that wander as well as a variety of other things, which can cause major problems.

☒ "Uniqueness"

Sometimes technical and facility risks combine to snag the unwary. Too often this is the direct result of the failure to consider all the factors that determine energy use in a facility… making assumptions without full understanding.

The varied conditions and system demands within some facilities can change efficiency "equations" in a manner that can lead to disaster. A modern hospital makes a good illustration. This kind of a facility, full of high tech equipment and very specific environmental requirements presents a wide variety of challenges and plenty of opportunities to get in trouble. Delicate and vital instruments require power quality far above most installations. Isolation areas will require negative pressures, operating rooms have rigid temperature and humidity requirements and the facility works on a 24 hour clock. Interruptions can't happen. Missing just one of these special requirements can destroy a carefully calculated project cost, cause lots of grief, and create undue suffering.

The key rule is: no two facilities, no two systems, are ever alike. A basic consideration and evaluation of the energy using systems within a facility is simply not good enough. Unless the auditor *understands* the specific facility and considers how the energy using systems relate to the whole facility, the audit will contain unacceptable levels of risk.

☒ *The Owner Insists*

There is one more area of technical risk that can cause problems. When a facility owner is enamored of a particular brand or type of equipment or system and insists that it be installed in the efficiency effort—a major risk may be waiting to happen. This raises the question of compatibility with what is already in place and may also limit the flexibility needed to obtain the best results. In the case of an in-house project, or one under the direction of an energy engineer, this risk may severely limit long term satisfaction with the project. For an ESCO working under a performance contract, the results can be very serious and require very specific mitigation efforts.

☑ **THE LAW OF AVERAGES**

Beware of the "Rule of Thumb." The problem inherent in applying average values, or "accepted" values, for energy use under various conditions is that conditions are seldom, if ever, *average*. The energy required by one piece equipment may be influenced by what is happening with another quite remote machine or system. Or, it may vary with how well the associated pieces of equipment are maintained. Or, how they are operated. "Average" disappears under observation.

There are rules of thumb that ascribe efficiency levels and/or loads to various kinds of systems. They make calculations easy, *but also wrong*. These values may be acceptable for "ball park" figures, but it will pay to remember that every ball park is different. Operating conditions are never identical and operators are never identical. If financial responsibility is riding on the accuracy of your numbers, values plucked from a table or rule of thumb estimates are just not good enough. If the "rule" says 270 sq. ft./ ton is right for hospital cooling, nod in agreement if you like, but *do a load analysis*.

If you have a table that gives the efficiency of various generic systems, use those numbers for "back of the envelope" calculations, but when there is money riding on your figures, measure the efficiency of the system you are dealing with.

A CLEAN SWEEP

This represents a very brief look at what can be a long list of potential risks that may affect a project. Experience will demonstrate the virtue of being inclusive in the risk inventory. Once assembled, it is easy to drop insignificant risks from the list. It can be costly, however, to overlook a risk that should have been on the list but later was conspicuous in its absence.

The key tool to apply in the identification of risks is an inquiring mind. Much of the process is based upon questions of how, why, why not and who, etc. These questions may have little to do with the technical side of energy efficiency, but everything to do with the results of an energy efficiency project.

These considerations are as important to the owners of a facility as they are to the energy engineers or ESCOs, who are anxious to put together a successful energy efficiency project. Owners have a major stake in the outcome and should play an active role in determining risk mitigation or management strategies regardless of whether an in-house project is contemplated or a full blown, comprehensive project involving a performance contract is in the works. Energy engineers and ESCOs clearly profit from a growing reputation for creating successful projects, which typically rests on predicted savings being consistently achieved.

Chapter 8

Potential
Mitigation Strategies

A CCORDING TO THE DICTIONARY, "mitigate" means to "make milder, less severe, less rigorous, less harmful." Nowhere does the dictionary suggest that mitigation means "elimination." Effective risk management, however, can sometimes reduce risk to a level of insignificance. For our purposes then, the process of risk mitigation means lowering risks where possible, providing options that may include acceptance of risks, management of risks or assignment of risks. In some cases of high risks or costly mitigation, it can mean rejection of certain measures and even abandonment of the project. These options provide the opportunity to assign cost values to risks and take those values into consideration when calculating the financial viability of a project.

Once the potential risks have been determined, it is time to review the list and decide what risks can be reduced to a "harmless" level, what can be managed, what can be accepted and what, if any, are "deal killers." Some will drop off the list as insignificant or be included within other categories. Others, taken in combination may add up to a cost level that potential savings cannot meet. Most, however, can be profitably mitigated if recognized, realistically evaluated and approached with some degree of innovation and/or ingenuity.

MITIGATING PEOPLE RISKS

Before we get down to specifics of risk mitigation, one point needs to be emphasized: The heart of risk mitigation, especially

119

when "people" risks are considered, is planned and continuous communications. What this boils down to is the establishment and *implementation* of a communications strategy, a plan that is in place and in operation long before the first hard hat is put on. Actually, even before there is a "project," communication is key.

Anything that can be done to increase the "buy in" by management, O&M staff or facility occupants will reduce project risk and enhance the likelihood of success. Without clear, frequent, open, two-way communications, the very best risk management ideas will, all too often, fail.

Strategies for mitigation can come in a variety of forms and with an even wider range of costs, but the question to be answered is, "Can this risk be rendered sufficiently harmless, in a cost-effective way, as far as the energy efficiency project is concerned?" The "how" depends on the nature and source of the risk and there can be no single (neat) strategic answer.

If, as suggested before, there is a solid energy policy in place but the policy has not penetrated deeply into the organization, the mitigation approach may be to encourage and assist facility management to communicate more effectively with their own people. This effort may require diplomacy on your part, but it can be well worth the effort. With surprising frequency, risks can be mitigated by simply getting people on the side of efficiency.

When it comes to mitigation, we cannot forget that the attitudes of the head of maintenance or the director of operations may present risks far beyond most others. All too often the suggestion that efficiency can be improved is perceived as a criticism of past practices, and often taken quite personally. It is worth repeating that if the head of maintenance or the supervisor of operations does not think a project will work, or doesn't want it to work, it will *not* work. The best, and only effective, mitigation strategy is to get these folks on your side by treating them with respect and getting across the idea that your project will help them with *their* job and make them look good. Give them some "authorship" by asking for their advice and listening. This is more than mitigation; they very likely will have good ideas.

And throughout the life of a project, "recognition" can continue to work wonders. Something as simple as a certificate presented in the presence of the boss will often wind up as a valued trophy on a boiler room wall. One ESCO took a maintenance crew to a football game in appreciation for their support for the energy efficiency effort. Small, low cost gestures that recognize performance can go a very long way to keep people factor risks at a manageable level.

In the occasional case of complete intransigence, someone in an influential position who is dead set against improvements; and will simply not work with the program, the selection of approaches is limited. Assassination is probably not an option. Which leaves us with the tricky option of appealing to our intransigent friend's boss. This takes some delicate maneuvering, but will sometimes work. Occasionally, however, the roadblock may be an old, valued employee or, much worse, somebody's relative. If that is the case there are two options, abandon the project, or make sufficient risk allowances to cover the possible (likely) problems that will result from the need to work around a key person. This can be very difficult and costly.

But, with some creative effort some roadblocks can be melted away. The brief case history below illustrates a risk mitigation strategy that cost almost nothing, and made a project much more successful.

Overcoming Obstacles; A Case History

"Prior to a facility survey at a local school district, we discovered the following biases held by the Director of Maintenance:

a. He didn't like T8 fluorescent fixture retrofit projects,
b. He didn't like locking thermostat covers or programmable controls,
c. He required custodians to turn on the HVAC system when they arrived at 4:30 a.m. (Why? Because at his last school the old steam boiler couldn't have the building warm enough when the students arrived if it wasn't turned on at 4:30 in the morning. The problem was, the school he now watched over had

packaged RTUs, which needed only about 20 minutes to bring the classrooms up to comfortable temperatures.)

The result? His school operated at an energy cost of $1.07 per square foot in a region where the average cost was $0.71 per square foot.

We had to make some subjective decisions at this point:

a. The Maintenance Director was going to be around his facility long after we were gone;

b. No matter how persuasive our arguments to his boss might be, this one person had the position and attitude to break the project;

c. We had a choice, walk away or try to bring the Maintenance Director into the project.

What did we do? We took the Director to four area schools where T8 lamps and electronic ballast had been successfully installed. He saw that the new lights worked without problems.

We *did not* include programmable controls for the HVAC system in our design for his school except to schedule On/Off operation. Time schedules were adjusted *every morning* for the first week of operation after the installation and the times most suitable to the Maintenance Director were used for final programming.

Let's review the process:

a. Recommendations were presented that would save the facility owner a significant amount of energy and expense.

b. Discussions with the key players were conducted and potential problems were uncovered.

c. The risks were weighed and determined to be a serious issue that could affect the success of the program.

d. A strategy was developed to minimize or mitigate the risks.

e. Revisions were introduced into the calculations and the resulting loss in projected savings was found to be acceptable, i.e., the owner's financial criteria were still attainable.

f. The revised program was implemented.

—Energy Systems Associates

This case history offers two important points: costs to the project in mitigating the risk... not only the cost of "courting" the Maintenance Director, but potential advantages in the installation of programmable controls were lost. The second point is the off-setting benefit achieved when the Maintenance Director became a part of the "action." He saw other systems in place and decided they would help him and *he* made the decision as to when heating started in the morning. He now had a stake in making it all work.

Risk mitigation can, and should, start long before a project gets underway... while it is still "a gleam in somebody's eye." This may not be a real formal endeavor... at least in the eyes of the occupants. To the O&M folks it might look like you are "just checking."

JUST CHECKING

One large ESCO begins the process informally on the very first tour through the facility of a potential customer. Their engineers go out of their way to talk to maintenance people, ask lots of questions, treat them with respect and lay the groundwork for later cooperation. The ESCO may not get the project, but if they do, they are well on the way to enlisting the maintenance staff as a real part of the action. If they get the job, they follow through with small performance awards and utilize other ways of giving credit to their client's employees who help make the whole project work. This way the maintenance folks feel an ownership in the project and have a stake in making it successful. A very low cost mitigation strategy, with good results.

But, what if the O&M situation needs serious improvement? What if maintenance plans are not in place, skill levels are low and overall conditions are less than great? The root cause may be well away from the control of the O&M supervisor's level. Solutions may require a mitigation effort at several levels. It may have to start at the very top if a solid energy efficiency policy is not in place and in use. If this is the case, mitigation might be very costly and/or difficult.

This is where the Master Plan, as discussed in Chapter 10, can come into play. This plan should cover (anticipate) maintenance requirements and the needed budget. But even if the plan exists, someone else may have laid hands on the budget, which unfortunately can leave the Maintenance Department with the short straw. In such circumstances, we have to look elsewhere for solutions.

One option could be to include an upgrading and expansion of the maintenance function as a part of the project. This may be a cost effective approach as there is a distinct probability that inadequate maintenance has left more room for efficiency improvements. This could be a place where a maintenance contract can be an effective mitigation tool.

Another strategy involves assisting the Maintenance Director in building (and presenting) the case for the improvements that are needed. If facts and figures can be assembled that will show management the benefits (profit) that can accrue from better maintenance, ways to make those improvements may be found. Remember, energy efficient O&M practices save energy and good maintenance can make a major contribution to IAQ. Often, the presentations can show that resulting efficiencies will more than pay the costs as well as offering a better work environment.

A major benefit of this approach is the "enlistment" of the Maintenance Director on the side of an energy efficiency project. Helping him do a better job, improving his life, makes him a stakeholder in the success of the project... a vital ally.

Mitigating the risk posed by low skill levels in the maintenance department is a different, but clearly related, problem. The solution, of course, is training. How much and how often depends upon current capabilities, skill levels required, the rate of staff turnover and how often "refreshing" is needed. It is a truism that well trained, highly skilled maintenance people are often lured away to a "better" job, leaving a skill gap in the maintenance shop. A gap that must be filled.

It may take some diplomacy to explain to the Maintenance Director that the staff lacks the skills necessary to make an energy efficiency project effective, but there is a good chance he already knows it. Presenting a way to upgrade the staff may help cement a relationship that will pay substantial dividends. The cost of bringing, and keeping, staff skill levels where they need to be is not negligible, but the cost of neglecting this important risk factor will be far more costly.

If the project is under a performance contract with an ESCO, risk assignment may be a way to handle training. Training costs may be accepted by the facility owner at a level and frequency agreed by both parties in the Energy Services Agreement. Most often, however, the responsibility for training is accepted by the ESCO and folded into the project financing package. This approach usually delivers cost-effective savings as well as mitigating major risks, for it provides the ESCO with somewhat more assurance that the measures that are key to the guaranteed results will be maintained at an appropriate level.

Aside from the direct advantages of the availability of skilled people to watch over efficiency improvements there is another human advantage that becomes very valuable. Well trained, skilled people tend to look beyond their instant jobs and will frequently find ways to do things better. Small improvements over the life of a project can add substantially to the results with no, or negligible, investment. A bonus to the mitigation strategy.

MITIGATING FACILITY RISKS

Effective mitigation of facility risks is dependent upon the care in which the nature, current use and future possibilities of the facility in question have been explored. If a facility is to be closed within the period of a proposed efficiency project, the mitigation strategy is simple. Walk away. However things are rarely that simple or well defined. Changes in the function or occupancy of a facility, changes in industrial processes or product lines, partial closing of a facility... a whole list of things that cannot be predicted with accuracy can change the economics of an energy efficiency effort. In some cases a change may increase energy use, in effect, negating the savings that were predicted. The opposite may occur with a cut back in hours of operation or other changes that reduce energy demand. The savings look better.

The problem, of course, is that either change can essentially wreck the economics of the project. If the demand drops, the project would get credit for improvements that it did not provide while, if demand increased through changes or expansion, the project could be accused of not delivering the promised savings. Either way someone is unhappy.

The mitigation strategy is two pronged. First, it is vital to anticipate changes as much as possible so that they can be factored into the project risk cushion. This may be a satisfactory approach if the likelihood of substantial change is remote.

The second strategy approach is much more formal and recognizes the role of a contract as a risk management tool. In a situation where a performance contract is in place, significant changes in energy demand resulting from facility or process changes are covered in the energy services agreement by a re-open clause. This clause provides that if demand changes from baseyear performance (up or down) by more than "X" amount, the baseyear shall be re-negotiated to account for the change. A somewhat more complex process, but an approach that can protect both parties to the agreement.

In the case of an "in-house" project it is important to identify energy use changes resulting from actions unrelated to the energy efficiency efforts so the effectiveness of the project can be fairly judged. The person, who sold management on the energy efficiency investment, should not be blamed when energy use goes up due to the addition of equipment or another shift of workers. Nor, should credit be given for "savings" when a wing of a building is closed.

If such changes are "in the wings," the potential energy consumption implications should be explored and communicated before the change occurs. Otherwise, a defensive voice may not be heard.

MITIGATING TECHNICAL RISKS

Excellent engineering capability can take precedence over some of the "art" in the mitigation of technical risks, even when challenges are posed by people or facilities. Of course, persuasion and diplomacy may well be needed as well.

The greatest technical risks are usually related to the integration of equipment designed to improve efficiency with the systems already in place. Tracing of the "energy chain" from the end back to the meters can identify potentially costly compatibility problems. Equipment and subsystems along that chain affect, and will be affected by, additions or changes resulting from an efficiency effort. The selection of equipment and systems, and careful development of the overall project design to successfully integrate with the chain offers effective risk mitigation.

What happens if a facility owner insists that his favorite brand of equipment or system should be included in a project. Unfortunately, all too often it doesn't really fit into the best way to improve efficiency or will not be fully compatible with systems already in place. The first mitigation strategy is the application of diplomacy and/or persuasion. This may be the only strategy available, particularly for an in-house project or one involving independent engineers. On occasion the "favorite brand" may be acceptable, but sometimes, definitely not.

The lure of "high tech" can pose a risk. Most of us like to think we are ready to adopt the shiny new ideas that our scientist and engineers come up with. Improvements are made every day and quite a lot of those improved pieces of equipment that hit the market work and work well. But some of them don't. The financing of energy efficiency projects is generally based upon improvements that produce returns over time; sometimes five or ten years. This means we have to have great confidence that systems and/or equipment designed to produce those returns will operate as intended over the life of the project. To be of value in an energy efficiency effort, they must have been proven to work.

The basic mitigation strategy for new technology is a simple rule: *Let someone else do the experimenting.* If guarantees, or your reputation, is riding on results, it is far better to install 5 year old technology that has been proven effective in a wide range of circumstances than to go for the latest stuff that looks great on the shelf.

The lure of high tech is world wide. In the '90s the authors assisted in the establishment of the first three energy service companies in China. As the companies selected names, each organization came up with a reference to high technology in their corporate name. We explained that this was a bad idea as they would be using '80s technology so they could guarantee results. Happily they changed their names and at this writing they are doing well.

If an ESCO is involved and guarantees are to apply, diplomacy and persuasion will play a role, but the risk management function of a contract also comes into play. Frequently, a joint selection procedure is used to select the equipment with specifications developed cooperatively followed by the assembling of a short list of vendors by one side with the other making the final selections. This allows some flexibility while still recognizing the need for the facility owner and the ESCO to both have a major role in equipment selection.

If there is serious concern about the viability of the project without ESCO control over equipment selection, guarantees may be eliminated, the work may go ahead with the ESCO acting simply as an energy service provider. The risk is thus assigned to the facility owner. If the concern over equipment is not too great, guarantees may still apply, but the "risk cushion" will widen to match the level of uncertainty. As a consequence, the size of the project in relation to investment will be reduced.

CALCULATING THE COST

It would be great if we had a formula into which you could plug a risk you have identified, fiddle with the numbers and bring forth a nice neat dollar figure for the cost of mitigation or management. Unfortunately, the process is just not that clean… just too many variables. But there are ways. If a good job of identifying and analyzing the risks has been done and the mitigating strategies clearly delineated, you are close to assigning numbers, not accurate to the last dollar, but sufficient to be included in the "risk cushion" assigned to the project.

It is at this point that the costs of various mitigating strategies must be weighed against their effectiveness in reducing risk. The very best mitigating option may be prohibitively expensive. In fact using a very costly option could rule out the implementation of the measure under consideration. A less effective, but less costly, strategy may be the best choice.

The costs of mitigation for most of the people issues, leaving aside the deal killing "stonewall" mentioned earlier, are generally not major considerations. A sound, well-executed communications strategy does not require much investment and, if done well, will bring benefits that go well beyond any cost involved.

In consideration of the O&M skill and/or personnel needs required to allow a successful project, the costs may be significant and will have an effect on the overall economics of the project. If the skills or manpower are lacking the cost, and risk, of not providing training becomes a huge concern. If basic maintenance procedures are lacking, the cost moves further up the scale. The question becomes, "with the maintenance structure the way I find it, how much will that detract from the likelihood of the project achieving the savings that should be there?" In other words, what will it cost to not make it right?

Determining the costs that may be posed by technical risks will fall more heavily on engineering expertise than on the "art." If the engineering aspect of an investment grade audit is well done and potential equipment or system problems have been identified *before the project gets underway*, the costs to meet these problems can be reasonably accurately calculated.

The cost for the risks associated with "high tech," as discussed earlier, are best settled through the negotiation of the energy services agreement. Risk sharing or risk assignment can be outlined in the contract.

There can be no hard and fast rules or formulas for attaching values to risk management costs. The best tool for controlling the cost is a thorough audit providing a true picture of the facility, its occupants and its energy systems as they are today, and a well considered assessment of what is apt to happen over the life of the project. This allows an organized analysis of the project as a whole and provides a basis for determining what percentage of potential savings can be realistically be achieved. Experience will improve the accuracy of this determination and it is quite realistic to expect a predictive consistency in the range of 90 to 110 percent of the savings potential in a project.

After the costs associated with a range of mitigating options have been established, it pays to sit down with the facility owner and "weigh" the options. Joint decisions at this point pave the way for acceptance of the project package later. This consultation, in, and of, itself, is usually a sound mitigation strategy.

Energy auditors have traditionally done some risk assessment and provided mitigation strategies based upon their experience and/or intuition. In many cases the results have been favorable, but in far too many instances, risks that were missed have lead to real problems. To help cover this, the ESCO industry, in the past, has rarely guaranteed more than 80 percent of the savings they believe can be achieved. A well done Investment Grade Audit considers risks in an organized fashion and provides a confidence level that has enabled some ESCOs to guarantee as much as 90 percent of the potential savings, making a real difference in project size. The key is the gathering of information... after all they call this the information age... and the intelligent use of that information.

Chapter 9

The IGA Report

ACH TIME an engineering firm receives a request for proposal (RFP) from a potential client, it seems almost inevitable that the information sought will include the question, "How long has your firm been doing energy audits?" The natural response of most engineering firms specializing in energy conservation and efficiency is to check back to see how long they have been in business and simply insert that number into the blank provided. Although that answer may be accurate, it is not really what the client wants to know. The real question should be, "How good is your firm at predicting the cost savings obtained through energy efficient installations and what hard data do you have to suggest that you actually know what you're doing?"

Now that question would be a bit more to the point and, quite honestly, more difficult than the normal *energy expert* would like to answer. So right now, before you respond to another RFP, ask yourself these questions:

1. What percentage of the energy audits prepared by your company *would you be proud* to submit as illustrations of your abilities?

2. What have you learned from your experience that has resulted in modifications to your report and/or calculation procedures that *distinguish you* from your competition?

3. Have your auditing skills improved over the years or are you still putting out the same old report with the same old assumptions and "rules of thumb?"

4.   Do you have records, i.e., *hard data*, demonstrating your skills
      in the key arenas of our business, including energy cost sav-
      ing projections and installation cost estimates. Have you
      checked back to see how well those estimates stood the test
      of time?

As someone who has been in the energy efficiency business
for over twenty-five years, Jim Brown's honest assessment is that
there are *very few* energy related firms who wouldn't squirm in
their chairs while answering these questions!

AUDITING HAS COME A LONG WAY

When the oil embargo hit our country in the early 1970's, we
were a pretty cocky breed. We had a president who believed that
we could relieve the foreign manipulation of our economy by is-
suing edicts to reduce room temperatures, drive slower, and travel
shorter distances. We had technical and environmental communi-
ties that believed the country needed only a few months of incon-
venience before some breakthrough would come along like solar
energy or the discovery of a new oil reserve.

Until that breakthrough could occur, we developed energy
analysis methods like the O&M Audits discussed in Chapter 1.
These audits consisted of checklists handed to maintenance departments for immediate 'quick fix' projects... and they worked!
Not because they were so brilliantly conceived, but because we
were such a wasteful nation. When we finally realized that the
energy shortage and high cost weren't just a passing problem, we
did a truly American thing... we developed an auditing procedure
so detailed and technical that it completely eliminated operations
and maintenance (O&M) from consideration. You had to spend
money to make money, and O&M projects just didn't cost enough!

We entered a time when: 1) we got tax breaks for installing
solar panels; 2) the expense of wind generator installations was
forced upon electric utilities by requiring them to buy all the
unused power that the system created; 3) performance building

codes were passed in several states establishing maximum Btu per square foot consumption levels. Then verification of these consumption ceilings by professional engineer's and architect's were required before construction could begin. These were "the good old days" for energy professionals. Even so, a great deal of good came to the nation as a result of the movement toward energy consciousness.

But, as time passed, oil began to flow freely again. The urgency to conserve diminished, and with it, the requirement for technical accuracy. We then entered a time of the "simplified" audit.

We got so good at this auditing stuff that we were able to tell you how long the payback period was going to be without calculating the energy savings... all we needed to know was the cost of the project, and we could tell you how many dollars you would save each year! Sort of a "No Calculation Required" investment.

Thankfully, (or should I say, hopefully), we have gotten away from these off-the-shelf energy analyses and are now producing more acceptable reports with greater emphasis on the balance between experience and technicality.

This movement back toward **reality** is occurring for two simple reasons ... consumers are more educated and energy investors are getting more sophisticated. A prediction of payback which misses the mark by one or two years simply isn't acceptable anymore. As a matter of fact, investments today require a degree of accuracy never before practiced within the energy industry. Predicted savings must now be so dependable that an energy service company (ESCO) can guarantee that a certain level of annual savings will be achieved, or the amount of the deficit will have to be paid to the facility owner.

That kind of financial arrangement produces a "reality check" of the highest order. Within months you will know if your calculations are right or wrong... there is no hiding, and very little hedging ...you are either right or you are wrong, and the consequences go straight to someone's pocketbook! Ultimately, yours!

That is why the movement is underway today to produce an

Investment Grade Audit. Investor's aren't prepared to write checks for inaccurate calculations, nor do they have time to wait for savings which may balance out "over the long run." They must have cost savings, and they must have them the first year ... the second ... and, the third.

What we are finding as we emphasize the *investment* in the audit procedure is that audit forms, calculation procedures, and technical expertise are not enough. Too many variables, many of which have been discussed in previous chapters of this book, fall outside the technical realm seriously impacting the annual bottom line for energy investors.

We are finding that maintenance staff *attitude* is just as important as *capability*; that management's commitment to energy savings is just as important as the *amount of money* they invest in the program; and, that the proper operation of energy related *ancillary equipment* (such as controls, valves and PE switches) is just as important as the time clock installed to turn equipment on and off.

There is increasingly an almost mind-boggling array of opportunities opening up for the truly *accurate* auditor. Opportunities like state agencies and public schools, which were once off limits to third party investment programs, today not only allow, but *require* guaranteed savings.

As a result, there will have to follow a shift in the energy consulting industry. A simple "trust me because I have a P.E. or C.E.M. at the end of my name" approach will no longer suffice. Energy professionals will be... are *being... required* to produce actual, measurable results. The initials at the end of the name, or the stamp on the front of the report may get you the first job with an investor, but results get the second!

Referring once again to Table 2-1 in Chapter 2, the Energy Systems Laboratory (ESL) analysis of the State of Texas' *LoanSTAR* program reviewed over 18 million square feet of building area, which had been provided almost $37 million for investment in 310 different energy measures and were calculated to save $9.6 million annually.

This table is a key assessment tool when attempting to determine how accurate the original estimates were... especially the *Measured Savings per Estimated Savings* column. Obviously, the goal is to be as close to 100% as possible, which would indicate that the estimated savings was right on the money. The actual range of predicted accuracy, however, varied from a Measured Savings 94.5% BELOW the estimate [an error percentage that would be unacceptable in any industry], up to 341.2% ABOVE the estimate [which, based upon what has already been explained in this book, is an equally egregious error]... the Average Measured Savings was 25% *LESS THAN* Estimated Savings.

That leads us to one of two conclusions: the engineers producing the reports couldn't spell Btu, or that these projects were running into some *unanticipated variables, which significantly impacted the results.*

Looking back at the table, the closest any package of recommended projects came to the achieving the estimated savings was 21.8%!

On the other hand, one of the graphics in ESL's analysis showing the Measured Savings as compared to the pre-retrofit annual utility costs depicted an average of 28% energy cost savings in the facilities surveyed. So, it's not that we don't know how to find energy waste ... it's just that we do a lousy job of estimating the value of those savings! Sometimes, we even seemed to have the graphics upside down.

The bottom line in ESL's report is summed up in a table depicting that the average project (a $487,162 investment) pre-

dicted savings of $128,525, or a 3.79-year simple payback period. Actual savings were only $96,240, with a simple payback of 5.06 years.

Here's the point... does a project that misses the estimated savings by 25% and the predicted payback period by 33% invoke enough confidence from the owner or investor to make him want to bring his checkbook to the next meeting you call? Is a 61-month payback good enough when they expected to have their investment back in 45 months?

Someone convinced these facility owners to borrow $11.5 million, and told them to expect an average annual return of $128,525. They got $96,240 each year.

WHERE HAVE WE GONE WRONG?

Let's look at a common calculation used for energy estimation and see if we can pick out a potential for error.

If we have a school campus with 200 4-lamp, 40-watt fluorescent fixtures, there is a total electrical lighting load in the range of 36.4 kW. If we desire to replace these lamps and the corresponding electromagnetic ballasts with 32-watt, T8 fluorescent lamps and electronic ballasts, the total load would be reduced to 20.3 kW.

Now, in an effort to produce an accurate analysis, the traditional auditor would talk to the "man in charge" (school principal, hospital administrator, etc.) to determine the operating hours of the lighting system and discover that they are on for 8 hours a day, 5 days a week for the entire year. So, he calculates energy savings as follows:

$$
\begin{aligned}
\text{Lighting Savings} \; &= \; (36.4 \text{ kW current} - 20.3 \text{ kW revised}) \\
&\quad \times (8 \text{ Hr/Day} \times 189 \text{ Days/School Yr}) \\
&= \; \underline{24{,}343 \text{ kWh/Yr}}
\end{aligned}
$$

If he had talked to the head custodian (whose job is to turn off all lights before he goes home each day), he would have found that the lights actually stay on 7-1/2 hours each day because his crew is out of there by 4:30 PM. The results…

Lighting Savings = (36.4 kW current – 20.3 kW revised)
                        × (7-1/2 Hr/Day × 189 Days/Yr)
                  = 22,833 kWh/Yr

The difference of 6.2% may not jeopardize the project's out-come too greatly, until he finds a box of lamps being used as re-placements, and the box contains 34-watt energy saver lamps! Now, he has to make a more thorough search of the existing light-ing system and discovers that one-half of the lamps are 40-watt and the other half is 34-watt. The new savings calculation is then:

Lighting Savings = (33.6 {ave} kW/fixture – 20.3 kW)
                        × (7-1/2 Hr/Day × 189 Days/Yr)
                  = 18,852 kW/Yr

Now, the error has grown to 22-1/2%. That's not quite as acceptable is it!

Did he use the right formula? Yes, he did. Is his methodology "technically" sound?

Yes, again. Then where did he go wrong?

He broke the most basic rules of *communication*. He asked the principal a precise question and assumed that the answer was equally precise. What he should have asked the principal was "Who would know how long the lights stay on?"

After the technical analysis confirms the opportunity, the next key is to step away from the chiller or the breaker panel and *really* look at:

a.   staffing ability to operate the system you want to install

b.   the reasons the systems are currently operated the way they are

c.   the ancillary equipment and it's impact on the retrofit project

d.   the plans for future facility scheduling and operation

In other words, technical calculations that justify our recommendations have been used over and over to convince investors to spend money... but, they should only be viewed as the "appetizer"... any money invested in a program that doesn't investigate the *subjective* and determine a strategy to handle the **real world** scenario is simply a financial disaster looking for a project!

It has become incumbent upon the energy professional to look into the *subjective side* of the situation. That is, if accuracy is truly desired.

Energy professional's have entered a time of *measured* performance.

Qualified predictions were once accepted as evidence for installation of specific projects, but not so anymore. Today, the owner wants guaranteed savings, and the investor looks to the energy professional to provide "guarantee-able" calculations and estimates.

The truth is that this is a call to *accountability* for our profession.

For us to step up to the plate and accept this accountability requires that we develop a means to become more responsible in our tasks... including the discovery process, the risk analysis process, the strategic implementation process, as well as the calculation and estimation process.

We must, of necessity, stretch beyond our normal... our traditional... methods and reach toward greater accuracy. So, let's look at the traditional to see what needs to be changed.

THE TRADITIONAL AUDIT REPORT

It is not difficult to find a book discussing the way to do a traditional audit these days. An impressive number of authors have published works describing their preferred methods for energy auditing and reporting. Almost every book and manual that has come across our desks has provided some new and interesting twists or concepts that could enhance an auditor's ability to develop technically accurate reports. For those who have read a few of these books, you would probably agree that there are as many different styles of reports as there are energy auditors. In fact, we could go so far as to say that many auditors use their report style as a competitive differentiator. However, as you analyze the available information, you will find that the basic format generally follows a pattern something like this:

- Pre-site visit tasks
- Site visit
- Post site visit tasks
- Report preparation
- Report presentation

Throughout this book, we have assumed that the reader is an experienced energy auditor with an established regimen for the collection of needed data. With this assumption, we now move directly into the report preparation phase.

It is further assumed that the client may be an owner or an ESCO. If the client is an ESCO, the auditor may find that the ESCO does not want all the IGA report conveyed to its client. In fact, the ESCO as a client may ask for a second modified report suitable for handing directly to its client.

In order to clarify the difference between the traditional and IGA reports, we need to select one of the myriad traditional report formats and pinpoint our recommended revisions. For purposes of our comparison, we have chosen to use the report format used by Albert Thumann and Bill Younger from their *Energy Auditing Seminar Workbook*[9-1]. It is an excellent format which produces an equally excellent point of reference... sort of a "springboard" into the IGA. As we progress through this chapter, we will suggest how this format for a traditional audit can be augmented to use as a format for an IGA. We, of course, expect that every auditor will then put his or her personal touch on their own IGA report form. Further, we expect that each report will be modified to reflect the client's particular needs and interests.

IGA REPORT PREPARATION

In the section-by-section description of a typical report provided below, the regular type will outline the material that would be in a traditional audit report. The material, which follows in each section that appears in *italics*, will be the suggested IGA additions to that section.

In your report, the risks analyzed and the associated mitigating strategies will often be implicit in the comments suggested in each italicized section noted below. We must warn you, however, that there is risk associated with pointing out risks. This is especially true when key decision makers have been intricately involved in the daily operation of the facility. As a result, you may find it prudent to discuss the risks with the greatest potential for disagreement or damaged relationships with the client prior to inserting the risk discussion into the report.

Maintenance is a prime example. If the equipment is not being adequately cared for, the professionalism of the maintenance staff will be suspect. So, the IGA auditor must become adept at describing the problem without alienating the Maintenance Department... a group that will obviously be key members of the project implementation team.

To summarize your mission (should you decide to accept it), the risk assessment will lead to recommended risk evaluation and mitigation strategies, which must be described without alienating those who will be key to the success of the program.

The italicized items below represent our best ideas on how to create an IGA and accomplish the desired results without shooting yourself in the foot along the way.

Executive Summary
Remembering that even though the IGA is a more complex

report, the report itself still needs to be as simple, straight forward and to the point as possible. That means that the first part of the Executive Summary should state the auditor's understanding of the basic "What you wanted us to do" and "What we did."

It will include: Introduction to the facility and/or process
                   Purpose of the audit
                   Overall conclusions

The IGA audit report will:
1.   *Add any significant circumstances or background information concerning the facilities which have a bearing on energy consumption.*

2.   *Discuss the methodology used in the selection of the final list of recommended projects, and the specific benefits to the owner. (Provide a summary narrative of the energy efficiency measures (EEMs) analyzed, EEMs not selected for recommendation (and why), and the overall benefits to the client.)*

3.   *Provide a summary narrative of each of the energy conservation measure (ECM) and/or EEM analyzed, including a brief assessment of the risk aspects, as appropriate.*

4.   *Conclude with a brief discussion of the "composite program" recommended. The sum total of installation costs and benefits obtained when all recommendations are implemented, taking into account the interaction of simultaneous operation of those recommendations; e.g., a lighting system renovation may reduce the building's cooling load and increase the building's heating load.*

(See Sample Table on following page)

Building Information
     This section should include the general background of the facility, mechanical systems and operations.
It will include: Envelope description
                   Floor plan
                   Operating schedule

Sample Table—Summary of Energy Cost Reduction Measures

SUMMARY OF PROGRAM

kWh Savings:	kWh/yr
Demand (kW) Savings:	kW/yr
Gas Savings:	mcf/yr
Btu Savings:	MMBtu/yr
Base Year Energy Reduction:	%
Cost Savings:	$/yr
Base Year Cost Reduction:	%
Implementation Cost:	$
Internal Rate of Return:	%
Net Present Value:	$
Payback Period:	Years

Occupancy patterns
Energy use in the plant
Mechanical systems
Operations and maintenance

The IGA report will:
1. *Add a detailed description of system controls including control types, temperature set points, boiler pressures, lighting controls, ventilation controls, calibration conditions and practices.*

2. *Add information regarding refrigerant types and air quality problems noticed, refrigerant monitors needed in equipment room, type and remaining allowable number of years for that refrigerant, etc.*

3. *Discuss system and design problems discovered that may be hindering efficient operation.*

4. *Discuss the suitability of the installed system for the purpose and/ or process served. (Has the usage of the area changed? Does the original system suit the revised use?)*

5. *Add any conditions found while on site, which might impact the decisions regarding project implementation, or those that may not have been known by the building owner and/or operators.*

6. *For more detailed renovation programs, create a spreadsheet with an Energy Consuming Equipment inventory that depicts nameplate data, loads, load factors, power factors, annual hours of operation, and current estimated consumption per equipment item. Balance anticipated consumption with actual historical utility bills comparing the sum total of estimated consumption to actual baseyear information to ensure that a credible "system overview" has been established.*

Utility Summary

Summarize the energy accounting you have used in the analysis (for the last two years if you have it); and, where possible, use graphics that are easy to understand.

It will include: Energy use index

                  Monthly consumption profiles

                  Demand profiles

                  Energy use history

The IGA Report will:

1. *Analyze utility expenses and develop, within the report, a calculation procedure demonstrating how utility bills are calculated by the utility supplier each month. This procedure should reflect real-world energy costs. Attach a copy of the current rate schedule in the appendix.*

2. *Include summary data from rate schedules establishing costs for each fuel unit saved by the recommend measures.*

3. *Discuss the current utility purchase process and available options to reduce energy cost.*

Sample: UTILITY RATE SCHEDULE ANALYSIS

ELECTRIC RATE SCHEDULE ANALYSIS
Utility Company:
Utility Company Representative for Client: Name ————————————
                                            Phone ————————————
Rate Schedule Analyzed:
Summary of Billing Component Charges:
Avoided Cost of Energy to be used in Savings Calculations:
Avoided Cost of Demand to be used in Savings Calculations:
Comments:
    EX: *(1)* *Periods of time when current schedule is most detrimental/beneficial,*
          (2) *Percentage of annual utility billing represented by electrical equipment usage,*
          (3) *Alternative rate schedules available to Client which may be more beneficial for*
               current operating practices,
          (4) *Cost per Btu of electrical energy.*

GAS RATE SCHEDULE ANALYSIS
Utility Company:
Utility Company Representative for Client: Name ————————————
                                            Phone ————————————
Rate Schedule Analyzed:
Summary of Billing Component Charges:
Avoided Cost of Energy to be used in Savings Calculations:
Comments:
    EX: *(1)* *Periods of time when current schedule is most detrimental/beneficial,*
          (2) *Percentage of annual utility billing represented by gas consuming equipment,*
          (3) *Alternative rate schedules available to Client which may be more beneficial for*
               current operating practices,
          (4) *Cost per Btu of gas energy.*

Energy Efficiency Measures (EEMs)[9-2]

Address energy conservation and efficiency recommendations and provide supporting calculations.

The EEM section will include:

Summary list of EEMs that meet financial criteria

EEM descriptions and calculations

EEM paybacks, with net present values, or life cycle cost analyses

EEM's considered that fell out of current financial criteria

In addition, the IGA report should provide:

1.   *A detailed budget for the installation of <u>each EEM</u>. The cost estimate should include audit fees, equipment, materials, labor, overhead, design, subcontractors, project management and administration, monitoring, financing, disposal, and contingency.*

2.   *The energy retrofit project implementation cost must consider an <u>owner</u> stipulated cost of capital for financial evaluation.*

3.   *Interaction of recommended EEM's.*

4.   *All assumptions clearly identified and carefully documented.*

5.   *Any environmental impact associated with the project.*

6.   *Selected M&V procedure*

7.   *Detailed commissioning plan*

8.   *Assessed risks and associated mitigating costs. List any unmanageable risks with a recommended approach, e.g., move risk into owner's area of responsibility or reduce in amount of guaranteed savings.*

Sample: PROJECT EEM TECHNICAL and
FINANCIAL ANALYSIS

SUMMARY OF ENERGY EFFICIENCY MEASURES (EEM) RECOMMENDED:

EEM #   EEM Name   Buildings/Systems Affected by EEM

All projects are to be analyzed in the dependent mode and in the following sequence:

Building Shell'Distribution Loads'Primary Equipment Loads'Energy Controls

ENERGY EFFICIENCY MEASURE:

EEM #: _____
EEM Name: _____

SUMMARY DATA:

Electric Consumption Savings: _____ kWh/Year
Electric Demand Savings: _____ kW/Year
Gas Consumption Savings: _____ mcf/Year
Other Fuel Savings: _____ ____/Year
Maintenance Savings (if applicable): _____ $/Year
Utility Cost Savings: _____ $/Year
Implementation Cost: _____ $
Payback Period (or ROI, etc.): _____ Years

EEM DESCRIPTION:
(Provide a narrative stating what the EEM will accomplish and how it is to be implemented. Describe the system operating hours, load, method of control, size and location. The analyst must keep in mind that the client must be able to read the EEM description and understand the logic for the measure. Include any clarifying sketches that may be necessary.)

ASSUMPTIONS:
(Summarize all assumptions which will affect the project implementation, cost estimates and cost savings.)

COST SAVING CALCULATIONS:
(Show the detailed energy cost saving calculations. Show all formulas, conversion factors and equations used to determine savings. All calculations must include units.
If computer programs are used, clearly identify all variables, columns of data and calculation procedures used to provide iterative calculations.)
[NOTE: Some clients prefer these calculations to be located in the report appendix.]

IMPLEMENTATION COSTS:

Material: _____ $_____
   Equipment: $_____
   Labor: $_____
   Subcontractors: $_____
   Report Preparation Fee: $_____
   Design Fee: $_____
   Administration/Project Management: $_____
   Monitoring & Verification: $_____
   Cost of Debt Service: $_____
   Contingency: $_____
TOTAL: $_____

(ESCO's may not wish to provide all of this information to their clients. However, all factors need to be considered by the ESCO prior to contract signing with the client.)

COST BACKUP:
(Provide unit pricing on all major items of equipment and material. Provide the Contractor estimates on all major installations that clearly break out material, equipment and labor. Include demolition and disposal costs.)

PAYBACK:
(Provide payback calculation using procedure requested by Client.)

NET PRESENT VALUE ANALYSIS:
(Optional financial analysis for project consideration.)

M&V PLAN:
(Compliance with IPMVP recommended.)

DETAILED COMMISSIONING PLAN:
(Describe how the entire program will be implemented. Include EEM priority, installation sequence, site preparation, demolition procedures, code compliance, permit application, environmental impact, system shutdown and/or bypass procedure, etc.)

RISK ASSESSMENT:
(Discuss perceived risks and recommended mitigation, i.e., increased cost, etc.)

An example of a typical Risk Assessment process may be seen in the following:

RISK ASSESSMENT

PROBLEM: Maintenance Staff is not able to properly service or maintain the desired system. No previous experience and/or training for this type system has ever been provided to current maintenance staff. No access to the required tools/equipment.

DECISION: Accept the task.

MITIGATING Add training to annual budget.
STRATEGY   Hire staff or contract with personnel experienced
or PLAN: with the systems.
                   Purchase necessary tools and equipment.

ACTION: Locate trainer or classes available for training.

COST: Revise budget to allow for expenses.

This example helps illustrate that the risk, the mitigation, etc., must be handled with care in the report—always keeping in mind that the remedy is to elicit the cooperation of the people related to the risk—not alienate them with imprudent observations.

Operation and Maintenance Measures
    Offer a brief description of specific low-cost operational improvements and maintenance considerations.

It will include: O&M recommendations
                    Cost and savings estimates

An IGA report will also:

1. *Discuss the recommendations for needed maintenance and operations revisions observed during the site visit.*

2. *Provide energy and energy cost savings calculations for each O&M.*

3. *Provide a list of Preventive Maintenance schedules obtained from equipment manufacturers.*

4. *Associated O&M risk mitigating considerations.*

5. *Significant problems eliminated or minimized through O&M practices.*

APPENDICES

The appendices will provide support material and technical information, including:

*   Floor plans and site notes
*   Photos
*   Audit data forms
*   Motor and lighting inventory
*   Rate schedules
*   Utility incentive programs, information and applications.

An IGA report will also:

1. *Include manufacturer literature for recommended equipment.*

2. *Provide schematic drawings of proposed system modifications.*

3. *Include proposals obtained for pricing purposes.*

4. *Introduce general system specifications for recommended equipment, if needed, for clarification of the recommendations.*

5. *Provide detailed calculations for recommended projects (if not included in the body of the report.)*

6. *Introduce Safety Issues Requiring Additional Analysis*

References

9-1   Thumann, Albert and William Younger. Energy *Auditing Seminar Workbook*. Association of Energy Engineers, 1996. Chapter 1 "The Auditing Process," Pages 1-7.

9-2   In keeping with our discussions in other parts of the book, we have selected the term "Energy Efficiency Measures" in lieu of the Thumann/Younger term "Energy Conservation Measures" to reflect that all measures may not "conserve" energy, and to recognize that efficiency puts the work environment first.

Chapter 10

Energy Master Planning: The Next Level

T HROUGHOUT THIS BOOK, we have called your attention to the need for IGA auditors to carefully consider many related issues, including the implications for indoor air quality. We have also declared that the IGA procedures are inextricably linked with measurement and verification. We have stressed that the auditor's clients should be provided with guidance regarding energy supply availability and matters of energy security. Further, we have underscored the clear relationship of the IGA to financing and, therefore, to implementation of a project.

These relationships suggest that any IGA takes into consideration many facets of an operation and the IGA itself takes on many roles and needs to reach into every corner of an enterprise. An analogy from Mr. William Shanner, paraphrased below, may help illustrate this point.[10-1]

Let's compare energy supplies to another bulk commodity, gasoline dispensed in a service station. We know how much gasoline is delivered to the service station. We also know how much gasoline is dispensed from each pump. We do not know the car the gasoline went into or what the car did with it... did it go anywhere? Was the trip necessary?... Did it get good mileage?... What is the total cost of transportation (operation, maintenance, insurance, finance, depreciation, parking, tolls, fees, etc.) of which gasoline is only a small part.

Similarly, until you know the total cost of the supplied energy as a function of the useful work it performs, you cannot make intelli-

gent decisions about energy's commodity cost, its effect on the environment, the services associated with it or how technology can affect its use. Most importantly, the auditor doesn't know how to best position the clients to compete effectively and grow in their respective markets.

The purpose of using Mr. Shanner's analogy is to emphasize that energy measures cannot be taken in isolation. As an IGA auditor becomes more sophisticated, the report will increasingly lay the foundation for a master plan. It seems particularly fitting, therefore, that the last chapter of this book talks about the ingredients of a master plan and how providing one allows the auditor to serve his or her client more effectively.

A key ingredient of an audit report is to maximize the ease with which the client can make use of the reported material. The best audit in the world does not save energy or improve the client's operation if it is not put to use. It seems an appropriate place to pull a quote from *Manual for Intelligent Energy Services.*[10-2]

The audit is a valuable tool, but audits don't save energy, people do! The unattended audit report gathers dust. Only when it is read, discussed, and implemented can its energy/environment/dollar benefits be realized. The difference between dust and energy savings is people. It is the communications connection that makes it work.

The kind of thinking put forth in the quote should become the conceptual backbone of an Energy Master Plan (EMP). The plan's strategy should integrate the audit's technical findings into the tasks and responsibilities of the people who must implement the plan. The whole program should be held together by a strong communications component—clearly set forth in the plan.

Perennially, the weakest two components in an energy management program are people and communications considerations. For a client to get the maximum benefit of an IGA, it is imperative that these two components be clearly spelled out. People and communications can make or break a plan that is *technically* perfect. Without solid communication that reaches key people and gains their support, it becomes only a question of who has the biggest rock, or the sharpest scissors.

A good EMP, however, is much more than managing energy; it is managing energy *implications* throughout the facility and/or processes. We are also talking about managing the investment guidance the audit offers to assure that the maximum benefit of the IGA is realized.

The Energy Master Plan proposed in this book is a tool to be used in the development of an integrated plan for the intelligent purchase, operation, upkeep and replacement of energy consuming and energy related equipment and systems. Although an EMP may be shaped to fit the needs of the particular owner or facility, its overriding focus should be the correction, renovation or replacement of particular energy consuming or energy related systems with the following priorities in mind:

- Life Safety Issues
☐ - Code/Standard Compliance
- Operating Expense   Decreasing
- Energy Efficiency   Priority
☐ - Equipment/System Age

You may be surprised to find energy efficiency so far down the list of priorities, but as the shock wears off it is hoped that you will begin to catch the underlying difference between the IGA and the Energy Master Plan. The IGA, by definition, focuses on projects that pay for themselves through energy cost reductions. Yes, there are ancillary issues that the good IGA auditor addresses (like IAQ, M&V, energy supply, and financial analysis), but the bottom-line is to install projects that *really* pay for themselves over time, taking risks and "what-ifs" into account wherever possible.

The EMP spreads the focus into other areas that affect the energy related systems, preparing a plan to operate safely, within accepted code requirements, while the *process toward efficiency and operational improvement* is accomplished. The EMP recognizes that it takes time to "get it all done" and that other related systems play as much part in occupant satisfaction as does energy.

Providing a master plan for the client is a value-added step beyond the IGA. In many instances it takes the IGA recommendations and tells the client how to get it done. If the IGA identifies a weakness, the master plan explores ways to ameliorate or remove that weakness.

An additional consideration is the potential long-term relationship with a client that is engendered through the creation and implementation of a master plan. The cost of customer acquisition is always a factor in running a business. To the extent you can better serve your client while offering more services, the more profitable the relationship will be. And over time, opportunities to sell even more services will present themselves.

Another global consideration before we talk specifics, is the effective, efficient use of energy. Too often audits and plans focus on energy reduction techniques. A master plan should respect the work environment and the total operation. The plan should assess the way energy can be secured and used most effectively—with the same, or even better, results.

Neither the plan nor the energy manager's job should be directed at conserving energy, but using the energy that must be used for an effective, productive work environment as *efficiently* as possible. One more time: energy conservation and energy efficiency are not synonymous—there are times when this difference can be tremendous.

Master plans can serve many purposes and some might not have anything to do with energy usage. They might, for example, be related to re-engineering an operation or to down-sizing the plant or the facility. The EMP that is addressed in this chapter, however, grows out of an IGA. A master plan might be as simple as a listing of the various organizational needs with assigned responsibilities, but typically a master plan growing out of an IGA will go beyond "to do" lists and address the interrelationships mentioned at the beginning of this chapter.

Even the most visionary auditor, energy engineer or energy manager cannot dream up a plan while sitting at his, or her, desk. The full value of the plan's potential interrelationships comes from listening, looking and gaining input from a broad range of people. Reality, be it a large bear, or a confused business office, must be part of the analysis. [Hint: see cartoon on the next page.]

The IGA auditor and those developing the EMP should view all facets of the client's operation through the lens of "energy availability": What is the source of energy for this process? How

effectively is it being used? How can its usage be improved without having a detrimental impact on the process? The answers to these questions cannot be derived in isolation.

An IGA auditor, who has effectively brought the people factor and risk considerations into his or her auditing procedures is in an ideal position to guide the master plan development.

DEVELOPING THE MASTER PLAN

Let's first look at the typical ingredients of a master plan and identify those pieces that are "boiler plate." Then, we can weigh various factors that might help us individualize the plan to particular conditions.

We fully recognize that not all Master Plans are created for energy efficiency. Some plans are developed simply to form a process for planned replacement of major equipment and/or facilities. For example, a school district that finds itself losing enrollment and income may face the need to maximize the use of its building space without investing huge amounts of money in facility remodeling. The "master plan" for that district may center around questions such as:

- Which campuses do we close?

- Which campuses do we renovate?

- What are the steps toward renovation and how will the needed work be prioritized?

Other plans may be developed to create a pathway toward the accomplishment of a specific future goal, such as a hospital's decision to add an additional medical service requiring highly technical and expensive equipment as well as the skilled technical staff.

Far too often we find replacement recommendations being made to owners prescribing the exact opposite order of priorities. Let us give you a couple of examples: When money is made avail-

FIRSTLY ... SECONDLY ...

able to replace a portion of the building's equipment, many turn automatically to the oldest equipment simply because it "should be" the next item to fail. This decision is frequently made because no one has gathered the data to really analyze the facts. Which piece of equipment has caused the most trouble over the last few months? Which unit costs the most to maintain? Is there equipment that is oversized... undersized... causes consistent comfort problems?

An Energy Master Plan would provide the information needed to make this decision based upon facts, not just gut-feel. The bottom line is that the older boiler, operating within acceptable efficiency ranges and within acceptable maintenance expense ranges should be replaced *after* the newer low efficiency rooftop unit that consistently loses compressors and never satisfies comfort requirements.

Or, we find equipment replacement budgets being spent for higher efficiency equipment just because it's higher efficiency equipment! The reasoning is sound. Increased efficiency produces energy cost savings that can, in turn, be used to replace non-energy saving equipment. But, when energy efficiency projects take precedence over safety issues, the repercussions may far exceed the benefits of the efficiency investment.

How many energy efficiency projects have you seen that were shelved because the cost to bring *ancillary systems* up to code was not included in the original project estimate? These are code/standard issues that simply cannot be ignored! Too often, project cost overruns occur because the project cost estimator did not account for the safety and/or code compliance issues that *must* be done if any work is allowed at all.

When in the process of EMP development, the plan developer needs to keep in mind that, unless unlimited resources are available, budgeted funds must be spent in order of priority. Not the auditor's priorities... the owner's priorities! And the owner's priorities incorporate the big picture, not just the currently recommended project!

Further, the EMP should evolve through planned communication strategies; so all perspectives are heard and considered.

Here is a good rule of thumb for the EMP developer:

"Always keep the Main Thing the Main Thing!"

A well-prepared EMP keeps the owner's priorities, life safety priorities, code/standard compliance priorities in focus at all times. At the same time, he/she prepares an EMP that blends *needed* projects so that they fit into the overall program right alongside the wanted projects.

Of all the benefits inherent within the Energy Master Plan, possibly the most beneficial is *Project Prioritization*. The goal of this portion of the plan is to assess future expenses, planning them into future budgets and reducing unexpected costs and budget-busting emergencies. Although no one can flawlessly predict the order of demise for individual equipment items, it has been found that given data such as age, degree of maintenance, hours of operation, etc., ones batting average, i.e., *predictive consistency*, can be significantly improved.

At this point, it would be good to discuss the differences, or rather the compatibility between the Energy Master Plan and Commissioning.

EMP and Commissioning

In a report entitled "A Practical Guide for Commissioning Existing Buildings," the staff at Portland Energy Conservation, Inc. and Oak Ridge National Laboratory define four different commissioning types or levels:

> "*Commissioning* is defined in ASHRAE Guideline 1-1996 as the process of ensuring that systems are designed, installed, functionally tested, and capable of being operated and maintained to perform in conformity with the design intent... commissioning begins with planning and includes design, construction, startup, acceptance, and training, and can be applied throughout the life of the building." The term com-

missioning, used in this sense, is generally accepted as an analysis of new construction projects.

Retrocommissioning (or existing building commissioning) is defined as "an event in the life of a building that applies a systematic investigation process for improving and optimizing a building's O&M." The report goes on to say that "its focus is usually on energy-using equipment such as mechanical, lighting, and related controls" and "is applied to buildings that have *not* previously been commissioned." Finally, the report states that "The retrocommissioning process most often focuses on the dynamic energy-using systems with the goal of reducing energy waste, obtaining energy cost savings for the owner, and identifying and fixing existing problems."

Continuous Commissioning is defined as a process much like Retrocommissioning with objectives that are essentially the same. The difference, however, is that Continuous Commissioning "more rigorously addresses the issue of persistence. A key goal is to ensure that the building systems remain optimized continuously."

Recommissioning "can occur only if a building was commissioned at some point in its life… recommissioning is a *periodic event* that *reapplies* the original commissioning tests in order to keep the building operating to design or current operating needs."

As you begin to develop your EMP, you will find that commissioning and the EMP's Engineering Survey go hand-in-hand, especially Retrocommissioning, Continuous Commissioning and Recommissioning. The EMP simply adds additional systems into the mix of those being analyzed.

Added EMP Components
Additional components that may be incorporated into the Energy Master Plan are:

Engineering Survey—An analysis of general system types, installations, problems and renovations needed for:

- Mechanical Systems: Heating, cooling, ventilation, refrigerant
  type, air and water distribution, domestic hot water, etc.

- Electrical Systems: From the transformer into the building,
  down to the power available in each building area. Electrical
  systems, like mechanical, wear out over time... they become
  overloaded as building loads increase... they should be analyzed and upgraded, even if no energy savings are to be
  obtained.

- Plumbing Systems: The types of fixtures, need for hot water,
  water conservation, ADA Compliance, odor sources—all
  need to be a part of the EMP.

- Structural Systems

- Information Technology Systems

- American Disabilities Act (ADA) Issues

- Indoor Air Quality Issues

A key goal is to classify the types of systems and estimate, as
closely as possible, the remaining useful life of the equipment.
During this task, an analysis of overall systems and the inter-rela-
tionship between systems is conducted.

Inventory of Systems—A specific analysis of individual equipment items including:

- Nameplate data: Serial number, electrical information, motor
  size... know what you have!

- Area served: know which system is causing the problem.

- Equipment age: estimate the length of life remaining.

- Equipment condition: what can be anticipated from this unit?

• Equipment maintenance history: what does it *really* cost to operate this unit?

• Suitability of the application for that equipment: is there something newer... better?

The EMP developer should dig out as much equipment history as possible. Maintenance records, operating expenses, frequency of problems, types of problems—all this information is needed when preparing recommendations for renovation programs. With this information, several of the questions raised while analyzing the overall systems can be answered. Why particular control strategies aren't working could be explained by the discovery of inoperable control valves or disconnected economizer damper linkages. This is where these problems are researched.

Another obvious benefit to this step is the accumulation of data describing the equipment being used. An inventory such as this can prove to be very beneficial when attempting to decide which equipment should be inserted into the budget two, three or even four years down the road.

Energy Analysis—This analysis serves to help the owner know precisely where his energy is being used and how much is consumed by specific items of equipment.

Energy Security—When the continuous presence of power is an absolute necessity, an analysis of the emergency generation equipment is needed. In addition, a study of the various types of fuel sources available, their dependability, and the potential for providing the same service through multiple fuel options becomes a high priority task. Alongside the security of the energy supply is its cost; however, when power is absolutely essential to an operation, availability will weigh more heavily than cost. Electric deregulation, utility contract negotiations, or simply explanations of the way the client is being billed for utilities can be included in this portion of an Energy Master Plan.

Operations & Maintenance Evaluations—Provide O&M evaluations for:

- Development of maintenance schedules for the equipment located during the inventory

- Preparation of man-hour estimates required to provide appropriate levels of maintenance

- Preparation of labor and parts costs for maintenance using recent parts billings, local area price proposals, owner supplied hourly rates and overhead costs.

- Creation of, or revision to, the maintenance department's work order system.

- Preparation of a preventive maintenance program for primary equipment items to improve operation and lengthen equipment lifespans.

Development of Software—designed to inventory and access individual equipment information including maintenance costs, parts numbers, operation recommendations, etc. This software can also be integrated into the work order system to track frequency of work provided and the degradation of equipment efficiency over time. It is also very beneficial during budget preparation to determine the specific costs of equipment scheduled for replacement in the next year.

Procurement Standardization—depicting lowest cost suppliers; standardizing the types of equipment and preferred components thereby reducing the number of systems that must be learned by the Maintenance Department and reducing the amount of inventory that must be stored. At the same time, this standardization can help insure consistent efficiency levels in new purchases.

Monitoring & Verification—M&V procedures needed in addition to reporting procedures required for effective communication of program results.

Training of Staff—and occupants regarding proper system operation, policy limits on temperature ranges and humidity levels, re-evaluation of operating hours for various kinds of equipment during various times of the day and year.

Common Components of an EMP
In broad terms, an Energy Master Plan should include:

* An assessment of current energy use and its implications for operation/mission and an inventory of the energy consuming equipment and systems;

* Existing and code operating parameters; e.g., air changes per hour (or cfm/occupant), temperature ranges, humidity levels, lighting levels, etc.;

* Baseyear information on what current consumption is and what conditions, such as occupancy, run times, operating hours, cause that consumption;

* An initial indication of energy efficiency savings potential (scoping audit) or complete recommendations from an IGA and specific organizational benefits which could result from such savings;

* Environmental concerns: indoor air quality issues, pollution and emission considerations;

* An assessment of supply options, including pricing, trends, and availability;

* Emergency preparedness and standby energy needs;

* An analysis of the operations and maintenance function; manpower, skills, training needs and related energy implications;

- Equipment replacement recommendations (including priorities by year);

- A determination of what needs can best be served by in-house staff and what tasks will need to be outsourced; and

- An Energy Policy.

Always a Work in Progress

An EMP should be a dynamic document ready to meet changes to the client's operation or external forces which may impact it. It will be most effective when it is flexible and responsive to potential changes. At a minimum, the EMP should be reviewed, and revised as appropriate, on an annual basis.

Whether the client is a school, hospital, industrial park, factory, or state government, the buildings are a collection of brick, limestone, wooden boards, and glass held together with glue, putty, mortar and nails. Sometimes they hold processes where widgets are combined with belts and motors to produce gizmos. No matter what labels are put on the edifices, products, or functions, the master plan needs to recognize that all the buildings and processes are part of the client's investment portfolio. The IGA identifies ways to increase the value of this portfolio. The master plan provides guidance in how to actually implement these recommendations as effectively as possible. With a long term view, such as a five-year plan, the document will also address ways to maintain and enhance these investments.

Any master plan will always be viewed by the various people in the client's organization through the perspective of their position in the organization and their responsibilities. The development of an effective plan will recognize the range of perspectives and the need for the affected people to understand what is being suggested. For example, recognition of a plan to enhance the organization's investment portfolio will appeal to top management and the financial people. It will do less for the facility managers and plant engineers, who are charged with developing and maintaining productive work environments.

Whether we are talking about running motors, turning on lights, conditioning air, or using energy as a raw material, intelligent use of energy resources requires management. Effective energy management requires knowledge and skills that seem to grow exponentially. As they grow, they require even greater support, authority and budget to make things happen. An absolutely essential part of any plan, therefore, is a statement of the energy, or resource, manager's responsibilities and the associated authority. It might also include an indication of the budget required to get the job done. As a result, we conclude this chapter with comments regarding the importance of an enforceable policy, designed to illustrate top management's commitment to the program.

The All-important Energy Policy

Although much has been written about energy policies and their importance to the successful implementation of an energy program, it needs to be stated *at least one more time* that nothing has more impact upon the success, or failure, of an energy program than this policy. If prepared correctly, the Energy Policy will ensure that the IGA report is used effectively and the Energy Master Plan will begin *and continue* as a credible influence upon the facility and its energy usage.

The Energy Policy's basic components are:

- Goal Statement: Specific, challenging yet attainable goals stated with the authority of the governing management.

- Plan Statement: The method(s) selected by administration to accomplish the goals, including responsibilities and authority of those implementing the policy.

- Specific Issues: A comprehensive list of issues that are to be resolved.

- Reporting: Frequency and content of reports required to ensure success of the program.

Although much more could be said about the Energy Policy, we conclude with the following:

The real issue faced by members of upper management when considering Energy Policy implementation is commitment... not occupant commitment, not staff commitment... but management's commitment! Are they committed to the success of the program, and are they willing to impart the authority necessary to assure the success of that program. Only through the publication of such a commitment, by the people who can really enforce it, can we ever expect to create a truly effective program.

And, only after the distribution of this administrative declaration can anyone be held responsible for producing the desired results. If given no authority to implement the program, the energy manager cannot be held responsible for the lack of results.

References
10-1 The analogy used here has been adapted from an unpublished work by William Shanner of RLE Technologies. 2003.

10-2 Hansen, Shirley J. *Manual for Intelligent Energy Services*. Published by The Fairmont Press, Lilburn, Georgia. 2002 pages 27-28.

Appendices

Appendices

Appendix A

M&V Options

THIS APPENDIX is offered for those who want more about measurement and verification (M&V) options than what is offered in Chapter 4; however, it does not provide sufficient information to perform M&V work. Anyone wishing to conduct actual M&V work should refer to the volumes published by the International Performance Measurement and Verification Protocol (IPMVP).

Options A, B, C, and D of the IPMVP are the basis of the established set of procedures. Options A and B focus on the performance of specific energy efficiency measures (EEMs). They involve measuring the energy use of systems affected by each EEM separate from that of the rest of the facility. Option C assesses the energy savings at the whole facility level. Option D relies on computer-based simulations of the energy performance equipment or whole facilities to enable determination of savings when baseyear or post-retrofit data are unreliable or unavailable.

Caution should be used in making comparisons with the results of Options A and B. They are intended to measure the energy use of each separate EEM. *The calculated savings from a series of measures are not additive.* Each time a measure is implemented the total energy consumption "pie" is reduced and the successive measure(s) will save proportionately less. If Options A and B are used to validate engineers' predictions which have been calculated in combination, the measures must be considered in the same order as the order used by the engineer in making his/her calculations.

Similarly, some measures have interactive potential with other measures. The classic example is the shift away from incandescent lamps. Over the years, we have described incandescent lamps as heat sources that just happened to give off light; so it is very obvious that changing out those lamps is going to affect the heating and cooling loads in a facility. When engineers calculate aggregate savings, they take this into consideration. Any M&V comparisons must follow the same procedure.

Thumbnail descriptions of the Options are offered below along with some application guidelines for each option. A little more attention is paid to Option A for several reasons: 1) it seems to be used the most and the 2001 revisions are important to its accurate usage; 2) most of Option B is rooted in A; and 3) Options C and D are typically used in larger projects and the end user is more apt to have an M&V specialist to assist in these endeavors.

In several instances the exact wording used in the IPMVP guidelines, often referred to as the MVP, are offered so you have a better sense of whether or not the contractor's firm is complying with the MVP when it claims it is.

THE OPTIONS

It's worth repeating: the descriptions presented below are not sufficient information for you or your colleagues to use the MVP. Those responsible for M&V for either party should read the full description of the options and suggested use in the Protocol and become familiar with their appropriate application.

Option A. Partially Measured Retrofit Isolation
Description

Savings are determined by partial field measurement (some, but not all, parameters may be stipulated) of the energy use of the system(s) to which an EEM is applied, separate from the energy use of the rest of the facility. Measurements may be either short-term or continuous. This option involves the isolation of the energy use of the equipment affected by an EEM from the rest of the facility.

Only partial measurement is used under Option A, with some parameter(s) being stipulated rather than measured. Such stipulations, however, can only be made where it can be shown that the combined impact of the plausible errors from all such stipulations will not significantly affect overall reported savings.

Stipulation may be based on historical data, such as recorded operating hours, indicated in the baseyear. Wherever a parameter is not measured in the facility for the baseyear or post-retrofit period, it should be treated as a stipulated value.

Factors to measure may be considered relative to the duties of a EEM contractor and the performance risks involved. Any factor that is significant in assessing the contractor's performance should be measured, while other factors beyond the contractor's control may be stipulated.

The decision as to which parameters can be stipulated rests on the significance of the impact of all such stipulations on the overall reported savings. Engineering estimates or mathematical modeling may be used to assess the significance of stipulation of any parameter in the reported savings. The M&V Plan should clearly state the stipulated values as well as analysis of their significance.

Initial measurements are critical. Measuring intervals center on the constancy of early measurement results. Section 3.4.1.4 of the MVP offers some guidance.

The MVP suggests sampling techniques that might be used, stating, "multiple versions of the same installation are included within the boundaries of a savings determination, statistically valid samples may be used as valid measurements of the total parameter." Appendix B of the Protocol outlines the acceptable sampling procedures and the level of uncertainty that exists. Be aware that the suggested procedures do not conform to scientific statistical procedures.

Estimates of energy savings using Option A with measured capacity can be adversely affected by:

• variation in operating efficiency;

- operational changes following measurements;
- malfunctions or lamp outages;
- related equipment changes; or
- failure to account for heating/cooling interaction.

Best Applications

Lighting retrofit where the power draw is measured periodically. The operating hours should be determined and accepted prior to installation. No one calculating the savings should accept the word of mouth version of hours of operation. Experience has shown that even in something with the routine operation of a school building, the principal has one version, the custodian has another figure and instrumentation almost always gives a third measurement. A Portable Data Logger is an inexpensive way to determine the hours that lights are on.

Option A is best applied where:

- performance of only the system affected by the EEM is of concern;

- interactive effects between EEMs or with other facility equipment can be measured or assumed to be insignificant;

- isolation of the EEMs from the rest of the facility and stipulation of key factors may avoid difficulty with non-routine baseline adjustments;

- independent variables affecting energy use are not complex and excessively difficult or expensive to monitor;

- sub-meters already exist to isolate energy use of systems; or added meters will be used for other purposes;

- uncertainties created by stipulations is acceptable to all parties;

- continued effectiveness of the EEM can be assessed by routine visual inspection of stipulated parameters; and/or

- stipulation of some parameters is less costly than measuring them.

Option B. Retrofit Isolation
Description
The savings determination techniques of Option B are identical to those of Option A except that no stipulations are allowed under B. Full measurement is required. Savings are determined by field measurement of the energy use of the systems to which the EEM is applied, separate from the energy use of the rest of the facility. Short-term or continuous measurements are taken throughout the post-retrofit. Such measurements may be taken on a periodic basis if acceptable to all parties involved. Continuous metering provides greater certainty in reported savings and more data about equipment operation. If these data are used to improve or optimize the operation of the equipment on a real-time basis, the added costs may be justified.

Best Applications
The savings created by most types of EEMs can be determined with Option B. Note that the application guidance below is very similar to Option A, but provisions regarding the use of stipulation are absent. Option B is best applied where:

- performance of only the systems affected by the EEM is of concern;

- interactive affects between EEMs or with other facility equipment can be measured or assumed to be insignificant;

- isolation may avoid difficult non-routine baseline adjustments due to future changes;

- independent variables affecting energy use are not complex and excessively difficult or expensive to monitor;

- sub-meters or other devices already exist to isolate energy use of systems;

- meters added can be used for other purposes;

- measurement of parameters is less costly than simulation in Option D.

Option C. Whole Building
Description

Option C is often referred to as the "Whole Building" approach; however, this option can be used for part of a building. It determines the collective savings of all EEMs applied to the part of the facility monitored by an energy meter. Short-term or continuous measurements are taken throughout the post-retrofit period. Option C usually relies on *continuous* measurement of whole-facility energy use and electric demand for a specific time before retrofit (baseyear), and *continuous* measurement of the whole-facility energy use and demand post-installation. Measurements may be taken on a periodic basis if acceptable to all parties involved.

Option C may be used in cases where there is a high degree of interaction between installed EEMs or between EEMs and the rest of the building, or the isolation and measurement of individual EEMs (as in Options A and B) is too difficult or too costly to use.

One caution in using Option C is warranted, the savings cannot be solely attributed to the EEMs. Other actions in the area may detract from, or add to, the appearance of the savings achieved by the EEMs.

Best Applications

Multiple energy efficiency measures are installed affecting more that one system in a building. Energy use is measured by the gas and electric utility meters for the baseyear (usually 12 months) and throughout the post-retrofit period. This "main meter" approach can make attribution of the savings to particular EEMs difficult in a dynamic situation where other factors are affecting consumption.

This option is intended for projects where savings are expected to be large enough to be discernible from the random or unexplained energy variations that normally found at the site. The baseline development generally requires regression analysis. The approach is sometimes referred to as "a system identification, parameter identification or inverse modeling approach." Certain assumptions are made and important parameters are identified through statistical analysis.

Option C is best applied where:

• the energy performance of the whole facility (or the area served by one meter) is being assessed; not just the individual EEMs.

• there are many different types of EEMs in one building;

• the EEMs involve diffuse activities, which cannot easily be isolated for the rest of the facility, such as operator training or wall and window upgrades;

• the savings are large enough to be separated from noise in the baseyear/baseline data during the time of monitoring;

• interactive effects between EEMs or with other facility equipment is substantial, which would make isolation techniques of Options A and B excessively complex;

• major future changes to the facility are not expected during the period of savings determination; and/or

• reasonable correlations can be found between energy use and other independent variables.

Option D. Calibrated Simulation
Description

Savings are determined through computer-based simulation of the energy use of components or the whole facility. Simulation routines must be calibrated so that they predict energy use and demand patterns that reasonably match actual energy consumption. Caution is warranted as this option typically requires considerable skill in calibrated simulation and data input can be quite costly.

Option D was added to the 1997 IPMVP to be used where calibrated simulations of the baseyear energy use and/or calibrated simulations of post-installation energy consumption can be used to measure savings. The simulations can be used for whole building or equipment subsystems analysis. This option may be used to confirm equipment performance, and may include one time or "snap shot" measurements of performance on an as-needed basis.

Major input variables that influence simulation results include:

- building plug and lighting loads;
- interior conditions;
- HVAC primary & secondary system characterizations;
- building ventilation and infiltration loads;
- building envelope & thermal mass characterization; and
- building occupant loads.

Be aware that there is no conclusive evidence that variables found to be important for one building will necessarily apply to another building.

Best Applications

When multiple EEMs are installed in a building that affect more than one system and no baseyear data are available, this option is preferred. Baseyear energy use is determined by simulation using a model calibrated by the post-retrofit period utility data. Skill required as well as the costs involved generally limit this option to large projects.

Option D is best applied where:

- either baseyear or post-retrofit energy data are unavailable or unreliable;

- there are too many EEMs to assess for Options A or B to be appropriate;

- the EEMs involve diffuse activities, which cannot easily be isolated from the rest of the facility;

- the impact of each EEM on its own is to be estimated within a multiple EEM project and the costs for A or B are excessive;

- interactive effects between EEMs or with other facility equipment is complex making isolation techniques excessively complex;

- major future changes to the facility are expected during the savings period;

- an experienced energy simulation professional is available and adequately funded for input and calibrating work; and

- the facility and the EEMs can be modeled by well-documented simulation software and reasonable calibration can be achieved against actual metered energy and demand data.

Appendix B

Sample Energy Policy

ENERGY MANAGEMENT POLICY
For

_____

School District

_____

(Date)

Mission Statement:
Recognizing our responsibility as Trustees of _____, we believe that every effort should be made to conserve energy and natural resources. We also believe that energy efficient operations will reduce operating costs and is in the best interest of the district. As a result, we are establishing this _Energy Management Policy_ which is to be implemented within each of our facilities and around all of our campuses. It is desired, through this policy, to produce a safe and productive environment for our students, while simultaneously providing prudent management of our financial and energy resources.

Statement of Concerns:
The district trustees are concerned about current and projected energy costs, the availability and procurement of electrical energy resulting from the deregulation of the electrical industry, and the power requirements facing the district due to current population growth patterns within the area. As a result, the development and implementation of a comprehensive, yet flexible, energy policy is believed to be in the best interest of the district.

Commitment to Implementation of Program:
Implementation of this policy shall be the joint responsibility of the trustees, administrators, staff and support personnel. The success of the policy is dependent upon total cooperation from every level within the system. Operation of the department shall be given oversight by the manager of the Energy Management Department.

Energy Management Department:
The Energy Management Department will develop a comprehensive program for energy efficient operation around the district. The goal of this program shall be to maximize energy efficiency throughout the district with proper consideration given to environmental and safety issues. The Energy Management Department will then be responsible for the implementation, operation and enforcement of the program. In addition, the department will:

1.) Evaluate energy rates and utility provider proposals to obtain the most reliable and cost effective energy sources available to the district.

2.) Routinely review efficiency improvements within pertinent industries and recommend new, more efficient equipment, systems and operating techniques.

3.) Work with campus principals and managers of other departments to develop an atmosphere of cooperation, and to establish acceptable operating practices among their staff and within their departmental practices.

4.) Annually review and revise these standard practices, as needed.

5.) Develop and promote educational energy awareness programs.

Energy Purchase—The Energy Management Department will be responsible for negotiations and purchase of energy required by the district for both current and projected future needs. Plans for the purchase, and distribution (if necessary), of energy for existing and planned campuses and facilities will be coordinated through the Energy Management Department.

Systems/Equipment Purchase—Minimum efficiency levels of each major system and equipment type shall be established by the Energy Management Department in cooperation with Construction, Maintenance and Purchasing Departments. In addition, the Energy Management Department shall assist these departments in the development of standardized specifications for energy consuming systems purchased by the district.

Operations—Specific operating practices of the district will be analyzed by the Energy Management Department and comments with recommendations will be provided to appropriate administrative departments and to an *Energy Committee*. This committee will consist of representatives from the Construction, Maintenance, Custodial and Purchasing Departments, along with representatives from the teaching staff, campus principals and district administration, and will be formed to establish enforceable rules and regulations to be followed under the energy efficiency program. This committee shall be chaired by the district Energy Manager. Decisions made by this committee will be presented to the school superintendent and board of trustees for approval and for authority to implement the specific recommendations. After acceptance of revised operations, notice will be filed with each effected department and the revision will be integrated into normal practice. Issues such as facility comfort levels, illumination levels, operating hours (facilities and equipment), community usage, after-hours activities and any other recommendation directed toward decreased energy costs shall be produced and recommended by the Energy Department after the approval process has been completed.

Education—The Energy Management Department will select an energy educator responsible for education of staff and students in the field of energy production, consumption and efficient operation. This educator will be responsible for communicating policy, distributing educational information about energy efficient operations relative to each specific campus, interpreting the success of the efficiency measures implemented, and providing the consistent stream of communication needed to keep energy efficiency as one of the major concerns of the district.

Reporting—The Energy Management Department shall produce monthly and annual reports providing actual consumption and energy costs for each district facility and/or campus. These reports shall provide comparisons of operating and cost requirements on a month-to-month and year-to-year basis. Reports depicting energy savings produced by energy efficient operations and/or renovation projects will also be provided, with success stories communicated throughout the district to increase awareness and involvement in the overall program. In addition, annual energy audits will be conducted at each campus to determine: facility additions/deletions, equipment/system operational revisions, alterations in primary facility usage, preventive maintenance revisions needed for improved operation of the aging equipment, revisions to facility inventory of energy consuming equipment, priority of equipment replacement, and any new or revised efficiency recommendations and/or practices available to each specific facility. An annual report summarizing these monthly and annual operating results and recommendations will be provided to the district trustees.

Having considered the responsibility of the district to conserve energy and to preserve our nations natural energy resources, improve the district's efficiency of operation, and eliminate unnecessary expenditures for energy, the _____ board of trustees does hereby adopt this *Energy Management Policy.*

Adopted this _____ **day of** _____, **2001.**

Signature: _____   **Attest:** _____
                                    President, Board of Trustees
                                    Secretary, Board of Trustees

ENERGY MANAGEMENT PROGRAM
for
_____

Energy Manager:
- Set up and implement the Energy Management Plan

- Establish and maintain energy records

- Identify assistance available from outside sources

- Assess future energy needs

- Oversee energy audits

- Identify sources of financing for energy projects

- Make energy related recommendations... purchase and efficiency

- Determine optimal balance between efficiency and safety/ health issues

- Assist construction, purchasing and maintenance departments in the implementation of recommendations

- Recommend operational revisions to improve efficiency

- Provide educational literature and programs for district staff and students

- Serve as chairperson of the Energy Committee.

Maintenance Department:
- Develop and fund a comprehensive Preventive Maintenance program

- Suggest situations when the maintenance department should have authority to change setpoints

Custodial Department:
- Recommend levels for custodial comfort during after-hours cleaning

- Suggest situations when custodial staff should have authority to change setpoints

Construction Department:
- Establish and enforce standard efficiency levels for primary equipment/systems

- Establish and enforce standard system types to be used within district facilities

- Create standardized specifications for energy consuming equipment

- Inform Energy Manager of codes, safety and health compliance issues that may impact the energy program and district operating efficiencies

- Inform designers of required district policies and standardized specifications

- Review and revise specifications and product requirements annually

- Research new systems and products used by other school districts annually

Purchasing Department:
- Obtain ordering and specification requirements from other departments

- Obtain list of "suitable" and "unsuitable" substitutes for products

- Inform potential bidders of requirement to follow district specifications

- Ensure that efficiency requirements are met when purchasing bulk items

Areas To Consider:
- Temperatures and humidity allowances for various areas

- Who has authority to revise scheduled hours of operation and under what conditions?

- Domestic Hot Water temperatures

- Operation of: kitchen equipment, classroom computers, and kilns

- Allowable time of day to turn on exterior lights, including activity fields

- Vending machine policy

- Portable heater and fan policy

- Value of ceiling fans in classrooms

- Value of ceiling insulation... problems with "conditioned" attic space... need for increased roof insulation levels.

- Community usage of facilities: areas and times allowed; cost for specific areas.

- If principals are to have final authority regarding after-hours usage at each campus, how will district handle public response to inconsistencies between campuses?

- Control of portable buildings

- *Accountability & Corrective Action*: How and to whom will violations of the rules be reported? What will be the procedure when violations of the program are discovered? If principals are to have final authority at each campus, how will excessive consumption be handled by the district?

- *Incentive Program*: Do you pass savings on to campuses? How much? What do you reward: Campuses with lower EUI than last year? Largest reward for campus with greatest improvement?

Index

*For Product Safety Concerns and Information please contact
our EU representative GPSR@taylorandfrancis.com Taylor & Francis
Verlag GmbH, Kaufingerstraße 24, 80331 München, Germany*

T - #0014 - 230425 - C0 - 234/156/11 [13] - CB - 9780824709280 - Gloss Lamination